TECHNOLOGY IN SOUTHEAST ASIAN HISTORY

Technology in Motion
Pamela O. Long and Asif Siddiqi, Series Editors

Published in cooperation with the Society for the History of Technology
(SHOT), the Technology in Motion series highlights the latest scholarship
on all aspects of the mutually constitutive relationship between technology
and society. Books focus on discrete thematic or geographic areas, covering
all periods of history from antiquity to the present around the globe. These
books synthesize recent scholarship on urgent topics in the history of
technology with a sensitivity to challenging perspectives and cutting-edge
analytical approaches. In combining historical and historiographical
approaches, the books serve both as scholarly works and as ideal entry
points for teaching at multiple levels.

TECHNOLOGY IN SOUTHEAST ASIAN HISTORY

Suzanne Moon

 JOHNS HOPKINS UNIVERSITY PRESS | Baltimore

Johns Hopkins University Press
2715 North Charles Street
Baltimore, Maryland 21218
www.press.jhu.edu

Library of Congress Cataloging-in-Publication Data

Names: Moon, Suzanne, 1962– author.
Title: Technology in Southeast Asian history / Suzanne Moon.
Description: Baltimore : Johns Hopkins University Press, 2023. | Series: Technology in
    motion | Includes bibliographical references and index.
Identifiers: LCCN 2022046838 | ISBN 9781421446912 (paperback) | ISBN 9781421446929
    (ebook)
Subjects: LCSH: Technology—Southeast Asia—History. | Southeast Asia—History.
Classification: LCC T27.S65 M66 2023 | DDC 609.59—dc23/eng/20221202
LC record available at https://lccn.loc.gov/2022046838

A catalog record for this book is available from the British Library.

*Special discounts are available for bulk purchases of this book. For more information,
please contact Special Sales at specialsales@jh.edu.*

*To the many scholars who have devoted themselves
to the study of Southeast Asia*

CONTENTS

TECHNOLOGY IN SOUTHEAST ASIAN HISTORY

Introduction

Technology in the Formation of a Region

THIS BOOK EXPLORES the history of technology in Southeast Asia, from ancient times until the early twentieth century. Technologies, including material artifacts and ways of making and doing, are embedded in every aspect of human life. Technological things, work, and knowledge, mediated by the influence of the physical environment, are vital for the dynamic production and reproduction of culture, society, power, and meaning. Whether taking advantage of seasonal floods to produce abundant rice harvests, building temples to reflect cosmological beliefs, or thwarting a trade monopoly by innovating in the production of competitive goods, creatively engaging with technologies does far more than help societies acquire the means of subsistence. Technological activities are important ways human societies create order, meaning, and social and political solidarity.

Sited at the intersection of the history of technology and Southeast Asian history, I offer a framework for understanding the meaning of technology in Southeast Asia over time. In recent years there has been a growing interest in the history of technology in Southeast Asia. Although rich seams of scholarship in both technology history and Southeast Asian studies are available to inform this work, attention to technology is fragmented by discipline and period. This book provides a

novel synthesis of scholarship about technology, highlighting important dynamics of technological change over time and across the region.
Historians of technology working on Southeast Asia have tended to
focus on modern histories, from the nineteenth century forward, drawn
to technology in colonial and postcolonial contexts. Although valuable,
this work can overplay the significance of incoming technologies and
underestimate or miss the continuities in technological culture and dynamics at work since prehistory. Scholarship in archaeology, anthropology, history, and other fields offer valuable insights into material aspects
of Southeast Asian life over its long history, although this material is not
always interpreted through the analytic of technology.[1] I address these
challenges by showing how profoundly entangled technology has been
in Southeast Asian politics, economics, and culture over its long history
and suggesting areas where more attention to technology could enrich
our studies of Southeast Asia.

The social and political diversity of Southeast Asia complicates this
task. How can we grapple with a region so diverse that some scholars
have questioned whether it can be meaningfully understood as a region
at all?[2] Rather than attempting encyclopedic coverage of technology in
the region, this book uses the theme of circulation to show how technology has informed the patterns of engagement and interrelated economic, political, and cultural changes that have constituted Southeast
Asia as a region.[3] Technologies facilitated and motivated engagement
between some societies and peoples more than others; they could aid or
diminish resilience and local sovereignty, support, transform, or un-
make local ways of life. Attending to circulation highlights important
continuities in Southeast Asia's technological dynamics, including long-
standing technological relationships and technological modes of engagement, and moments of true rupture. It can help us dismantle polarizing narratives of technology as either simplistically "indigenous"
or "foreign" by exploring technological dynamism over long historical
periods and drawing our attention to processes of innovation, adaptation, and domestication that have embedded technologies firmly (or less
so) in the social fabric of Southeast Asian societies.[4] Finally, the theme
of circulation offers insight into technology's role in creating, maintain-

ing, or reducing cultural or social diversity through technologically mediated economic interactions, migrations, and warfare.

Southeast Asia: An Introduction to the Region

Because Southeast Asia's diversity plays such a critical role in this book, a brief introduction can offer useful context, especially for those unfamiliar with the region. In the contemporary world, Southeast Asia consists of eleven nations (figure 1). Scholars often distinguish between the mainland Southeast Asian states of Myanmar, Thailand, Laos, Cambodia, and Vietnam, and the island states of Malaysia, Singapore, the

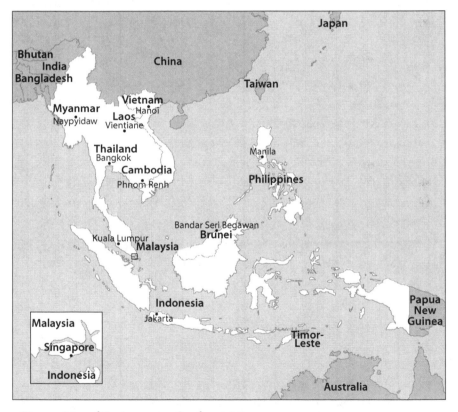

Figure 1. Map of Contemporary Southeast Asia.
Courtesy ASEAN UP, https://aseanup.com/wp-content/uploads/2016/11/ASEAN-map.jpg.
Public domain.

Philippines, Indonesia, Brunei, and Timor-Leste. The region stretches for nearly 4,500,000 km², with more than 4,340,000 km² of land.⁵ The climate varies from the distinct dry and wet monsoon patterns of the intertropical zone of the mainland and the Philippines and the year-round rainfall of the equatorial zone.⁶ Historically, the tropical climate supported rain forest cover in much of the region. Deforestation, both recent and in the more distant past, has shrunk the areas of both temperate and tropical rain forests considerably.

Although sparsely populated for much of its history, Southeast Asia is now home to 8.6% of the world's population (approximately 670 million people).⁷ Indonesia and Thailand are the largest economies in Southeast Asia by gross domestic product (GDP), with agriculture, trade, and manufacturing supplying a large percentage of their incomes.⁸ Although agriculture is important in nearly every country, there is significant variability in economic focus.⁹ Since 1967, the Association of Southeast Asian Nations has promoted economic and diplomatic integration between nations and supported efforts to grow and diversify Southeast Asian economies.

The following sections offer a schematic introduction to the geographical, religious, and linguistic character of the modern nations of Southeast Asia. Although hardly a comprehensive picture of diversity, these three areas suggest the multiple distinctive cultural lifeways that coexist even within nations and important transnational patterns of cultural connection. Linguistic evidence in particular highlights deep histories of connection and splintering among ethnic groups over time.¹⁰ Together they offer a basic sketch of cultural diversity across this vast region.

Myanmar, Thailand, and Yunnan

Located in the extreme west of Southeast Asia, Myanmar is the largest country on the mainland (figure 2).¹¹ Its long coastline on the Bay of Bengal and the Andaman Sea historically facilitated trade and close connections with the Indian subcontinent. The Ayeyarwady, Salween, and Mekong Rivers, running north to south, offered easy movement away

Figure 2. Map of Myanmar, Thailand, and Yunnan.
Courtesy ASEAN UP, https://aseanup.com/wp-content/uploads/2016/11/ASEAN-map.jpg. Public domain.

from the coasts and valuable water resources for settlement and agriculture in the rich alluvial plains, especially rice agriculture. By contrast, people in the less watered upland areas practiced less intensive (but highly sustainable) shifting cultivation—an approach in which farmers rotate fields rather than crops, leaving some lands fallow to recover fertility over time.[12] Inland temperate and tropical rainforests supplied daily needs and valuable products for trade in historic times.[13]

Thailand also has broad fertile floodplains, dry uplands, and forests. The central plain formed by the Chao Phraya River has supported agriculture since prehistoric times. Just as in Myanmar, farmers in upland regions of Thailand employed different forms of agriculture than those working the fertile plains. The long Malay peninsula has given the peoples of Thailand access to maritime trade via both the Andaman Sea in the west and the Gulf of Thailand in the east.[14]

The inclusion of Yunnan, now a province of southwestern China, in Southeast Asia is justified by Yunnan's deep history. In the early periods covered in this book, the peoples of Yunnan were closer in culture, language, and social practices to those in Myanmar, Thailand, and Laos than to China. Their upland agriculturists used similar techniques as those in Myanmar and Thailand, and they shared the Salween and Mekong Rivers with Myanmar (rising in the Himalaya mountains farther north). Thus, at least until the modern period, including Yunnan helps us see more clearly the full regional dynamics of Southeast Asia.

Speakers from five major language families still live in mainland Southeast Asia: Tai-Kadai, Sino-Tibetan, Hmong-Mien (also known as Miao-Yao), Austroasiatic, and Austronesian.[15] Aside from wide use of the national or state language, each country has multiple living languages that hint at the real social diversity and transnational connections obscured by national borders. Myanmar's official language is the Tibetan-Burman language of Burmese (in the Sino-Tibetan family), but at least 100 other languages are also spoken.[16] In Thailand, Tai languages from the Tai-Kadai family are most prominent, including the national language, Thai.[17] Yet sizeable minorities speak Sino-Tibetan or Austroasiatic languages. On the border between Myanmar and Thailand are numerous speakers of three mutually unintelligible Karen languages, all from the Tibeto-Burman family. Shan, a Tai-Kadai language, is spoken in the Shan state in Myanmar and parts of northern Thailand. Mon-Khmer languages, a subgroup of the Austroasiatic family, have numerous speakers in Thailand.[18] During the periods covered by this book, the people of Yunnan were mainly speakers of Tai-Kadai languages.[19] Notably, while official or state languages on the mainland tend to be drawn from the common languages of the alluvial plains, many minority languages are spoken by peoples in upland regions.

In religious practice, both Myanmar and Thailand report the majority of their population as Buddhists, with minorities practicing Islam and folk religions. Chinese religious traditions dominate in modern Yunnan, although there are also significant minorities of Buddhists. Buddhism was the most widely observed religion among the diverse Tai peoples of the region in the periods covered in this book.[20] In each of these countries, the Theravada branch of Buddhism was most influential, arriving directly from the Indian subcontinent.

Laos and Cambodia are bordering countries located on the Indochina Peninsula (figure 3). The western border of Laos is the Mekong River, a major conduit for trade and travel in the wider region. The rest

Figure 3. Map of Thailand, Cambodia, Laos, and Vietnam.

Courtesy ASEAN UP, https://aseanup.com/wp-content /uploads/2016/11/ASEAN-map.jpg. Public domain.

of the country is mountainous. Despite deforestation, it still has both evergreen and deciduous rain forest cover.[21] Laos is landlocked, while Cambodia borders the Gulf of Thailand. The Mekong and the Tonlé Sap River and lake provided vital water resources that fueled the development of intensive agriculture in the fertile central plain. Like those in Myanmar, Thailand, and Laos, Cambodia's forest resources have suffered considerably from overexploitation in recent years.[22]

In Cambodia, the national language is Khmer, an Austroasiatic language of the Mon-Khmer family. Minority languages include, among others, the Austronesian language Cham, which was the primary language of the kingdom of Champa (192–1832 CE).[23] Laotian, the national language of Laos, is from the Tai-Kadai family and shares some vocabulary with Thai. However, even in this relatively small country at least 80 minority languages are spoken, including Hmong languages (of the Hmong-Mien family).[24] Like Thailand, Myanmar, and historical Yunnan, both countries report majority Buddhist populations, being similarly influenced by the cultures of South Asia. Folk religions are also widely practiced. In Cambodia, official numbers suggest Buddhism is practiced by 98% of the population.

Vietnam

Vietnam's history has been shaped by its long coastlines, mountainous regions, and major river systems, including the Red River, which rises in Yunnan and flows into the Gulf of Tonkin, and the Mekong, which empties through the Mekong Delta in the southwest of the country (see figure 3). Both maritime and agricultural economies played a significant role in the rise and fall of Vietnamese empires.[25] The national language of Vietnam, Vietnamese, is an Austroasiatic language. Minority languages include Tày, a Tai language spoken in the borderland regions with China, Khmer, Cham, and several dialects of Hmong. As in other parts of the mainland, clusters of minority speakers are often found in the hill regions.

Vietnam's relationship with China over more than 2,000 years has significantly shaped Vietnamese culture and governance. China invaded northern areas of Vietnam four times before 1000 CE, resulting in numerous appropriations of Chinese culture by Vietnamese leaders and ordinary people. Most Vietnamese indicate no religious affiliation or affiliation with Vietnamese folk religion, which is strongly influenced by Chinese Daoism and Confucianism. Large minorities follow the Mahayana Buddhist traditions common in East Asia (China, Japan, and Korea).

Malaysia, Indonesia, Singapore, Brunei, and Timor-Leste

The Malay Archipelago, made up of thousands of islands, with 17,000 islands in Indonesia alone, is home to five Southeast Asian countries (figure 4). Given its geography, it is no surprise that maritime-oriented societies played a prominent role in the region's history. However, large islands like Borneo, Sumatra, Sulawesi, New Guinea, and Java and smaller islands like Bali and Timor have also supported polities heavily dependent on agriculture and foraging. Volcanic activity is common, operating as both a curse and a blessing. Although volcanoes have undoubtedly disrupted human occupation of the region, they have also frequently provided remarkable soil fertility that attracted settlement. Dense rain forests covered much of the archipelago in the past. As on the mainland, contemporary deforestation has made serious inroads in nearly every country except Brunei.[26]

There are hundreds of languages spoken across the Malay Archipelago. In Indonesia alone, there are more than 700 living languages; in

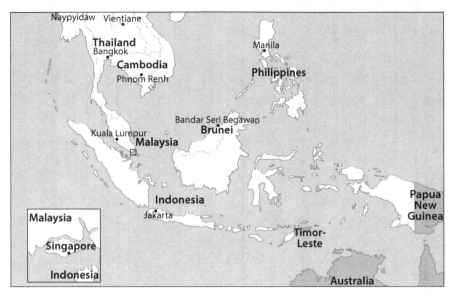

Figure 4. Island Southeast Asia.
Courtesy ASEAN UP, https://aseanup.com/wp-content/uploads/2016/11/ASEAN-map.jpg.
Public domain.

Malaysia, there are 137. Austronesian languages predominate around the archipelago, including the related national languages of Indonesia, Brunei, Malaysia, and Singapore. An admixture of Mon-Khmer languages are spoken in Malaysia, particularly on the peninsula shared with Thailand. Papuan languages—the largest number of non-Austronesian languages in island Southeast Asia—are spoken in New Guinea and other nearby islands.[27] Timor-Leste has two official languages: Portuguese and Tetum.

Malaysia, Indonesia, and Brunei, which account for the vast majority of the population of the archipelago, are majority Islamic nations, with significant minorities of Christians, Buddhists, and practitioners of folk religions, as well as somewhat smaller minorities of Hindus in both Malaysia and Indonesia. Timor-Leste is a majority Roman Catholic nation. The tiny island nation of Singapore records practitioners of nearly 400 different religions thanks to its highly diverse population, with Buddhism practiced by a small majority.

The Philippines

The Philippine Archipelago comprises more than 7,000 islands, although not all are habitable. Like much of the Malay Archipelago, the Philippine islands are volcanic (see figure 4). The largest and most densely populated island is Luzon, but Mindanao, Negros, Panay, Cebu, Leyte, and Mindoro also have significant populations. Agriculture is a mainstay of the economy. Like many other nations in Southeast Asia, the Philippines was formerly heavily forested but has lost most of its forest cover in recent years, with just a few areas of primary rain forest left.[28] In the southwestern part of the country, the Sulu Archipelago has been a vital player in regional maritime trade for much of its history. It has been a source of desirable ocean products such as tortoiseshell and physically linked the Philippine and Malay Archipelagos.

There are around 200 languages spoken in the Philippines, with Filipino, the official version of Tagalog, as the national language. The indigenous languages of the Philippines are all Austronesian languages. The Philippines is majority Christian, with Catholics more prominent than Protestants. It has a sizeable minority of Muslims and practitioners of folk religions.

Plan of the Book

I explore Southeast Asia's technology from prehistory until the early twentieth century. Since ancient times, kingdoms, sultanates, empires, and population centers emerged in different parts of the region, creating a dizzying array of states, empires, and centers of trade. Cross-border dynamics are also important, especially in the areas between Southeast Asia and China. I balance the need for broad coverage of a diverse region and a desire for clarity by using telling and well-researched historical cases to draw attention to widely significant patterns and processes of technological change.

The first chapter explores the emerging technological cultures—settled, nomadic, and seminomadic—that characterized human life in prehistoric times and antiquity. The mutual shaping of technology and forms of social, political, and economic interconnection, including growing interdependence between different peoples and polities, is the core focus. Technologies facilitated particular forms of social and political engagement, and increasing circulation and exchange likewise shifted technological possibilities, aims, and ambitions.

The second chapter explores technology, agriculture, and the emergence of urban societies around the region, from early settlements to the large cities that anchored major Southeast Asian kingdoms into the fourteenth century. Cities were centers of innovation in construction and crafts, nodes for expanding trade, and motivated intensification of agriculture. They integrated technologies in remarkably durable and effective ways. Their technical strategies supported dense human settlements that grew and thrived.

Chapters 3, 4, and 5 explore technology in the explosion of commercial and political change from the fourteenth century until the early nineteenth century. This period saw expansion and reconfiguration of trade and significant political and economic change for many societies across the region. Each chapter explores a different aspect of technological dynamism that played a significant role in the region's wider social and political transformation. Chapter 3 uses the case of textiles to explore the technological character of Southeast Asia's commercial culture and the disruptions caused by growing demand in both Europe and China for Southeast Asian products. It offers insight into the role of technology and technological learning in the resilience of Southeast Asian societies during these tumultuous years. Two cases explored in chapter 4—the expansion of labor-intensive Chinese mining in Southeast Asia and hybridization of local and foreign ideas in shipbuilding—shed light on how imported techniques and technologies shaped the early modern period. Chapter 5 investigates the expansion and intensification of agriculture, which served as a foundation for the later emergence of export-oriented agriculture in European imperialism.

Chapters 6 and 7 explore technology and warfare. Chapter 6 considers the technological foundations of conflict that stretch back to antiquity, and the changes and continuities in technical and military practices that emerge in the early modern period. It investigates the introduction of guns and the tactics and strategies employed by foreigners to understand how new military technologies and techniques were woven into preexisting forms of conflict. This focus on integration highlights how technologies of warfare shaped patterns of human circulation around the region and helps reveal how the reality of violent conflict affected the lives of Southeast Asian peoples. Chapter 7 considers the diverse sociotechnical approaches that Southeast Asians used when the scale of warfare expanded to larger armies, and over larger areas.

Chapter 8 investigates technological change in the nineteenth and early twentieth centuries, focusing on the integration of Southeast Asian modes of work and production with incoming technologies, and considers Southeast Asian social and cultural responses to the dramatic changes they saw in this period.

CHAPTER 1

Technology in the Human Settlement of Southeast Asia

ISTORIANS, ARCHAEOLOGISTS, AND ANTHROPOLOGISTS have often referred to Southeast Asia as a crossroads region.[1] The crossroads metaphor captures Southeast Asia's historical role as a valued site for trade and migration and implicitly acknowledges the diverse cultures that emerged there over time. This chapter explores technology in the human settlement of Southeast Asia. Early migrants created within Southeast Asian environments a mosaic of nomadic and settled ways of life, developing diverse sociotechnical strategies of subsistence and flourishing.[2] Technologies and characteristic itineraries of movement were mutually constructed over time in response to the affordances and challenges of Southeast Asian environments. Highlighting the relationship between technological change and mobility offers valuable insight into both the survival and sustainability of human life in the region and the diverse ways of life that came to characterize the region. This diversity played a significant role in making Southeast Asia the attractive destination for trade that it ultimately became by 200 BCE.

The First Wave of Migration into Southeast Asia: Hunter-Gatherers

Archaeologists have provided persuasive evidence that Southeast Asia was populated via two waves of migration by anatomically modern humans (*Homo sapiens*).[3] The first migrants were hunter-gatherers, groups who subsided mainly on nondomesticated plants and animals. They began migrating into the region no later than 47,000–50,000 years ago (figure 5).[4] Their most likely path started in Africa and from there moved into tropical Eurasia and then mainland and island Southeast Asia over the course of millennia.[5] Although there is some debate on this point, physical anthropological evidence suggests identifiable biological differences between the first wave of migrants and later Neolithic populations.[6] Using DNA and craniometry, anthropologists link the first wave of migrants to contemporary communities and skeletal remains of people native to the Andaman Islands, Australia, Papua, and parts of Japan, a group sometimes designated as "Australo-Papuan."[7]

The technologies left behind by these Paleolithic populations suggest a fully nomadic lifestyle; typical sites of human occupation in which tools like axes or spear points are found appear to have been occupied only intermittently. They engaged in broad-spectrum subsistence practices, hunting diverse animals and foraging for a wide variety of plants.[8] For roughly 48,000 years, hunting and foraging populations dispersed throughout the coastal areas and forested interiors of mainland and island Southeast Asia, wherever ecologies were rich enough to support human life. The population density varied with local ecologies, yet this way of life proved remarkably adaptable and resilient. Even areas like the equatorial rain forests of central Borneo, which lacked sufficient edible plants or animals for any dense population to grow, saw occupation by small groups going back at least 50,000 years.[9]

Although much migration would have happened on foot, there is good evidence of the use of watercraft from very early times.[10] Sea crossings out of sight of land took place as early as 45,000 years ago between the eastern areas of the Malay Archipelago and northern Australia. The large island of Sulawesi (east of Borneo, now part of Indonesia), acces-

Figure 5. The "Bouquet of Hands," an example of Borneo rock art, features six hands. Note the older orange-red hands (*light-gray* in the photo) covered by the motif. Found in the Jeriji Saleh Caves in East Kalimantan (Borneo), Indonesia. It is estimated to be 40,000 years old.

sible only by the sea, was settled no later than 28,000–30,000 years ago, with indisputable evidence of water crossings to neighboring islands at least 100 km away.[11] Precise knowledge about the technical characteristics of the watercraft used at these early dates is not available. Most were probably made from nondurable organic materials like wood and bark. But skeletal remains, extant tools and ceremonial objects, and the DNA profiles of current populations strongly support the existence of paleolithic sea crossings and seaworthy craft.[12] From around 2000 BCE, more indirect evidence confirms that people used boats suitable for the deep ocean. For example, findings at the Nong Nor archaeological site in coastal Thailand show that inhabitants ate bull sharks and eagle rays, which are only caught in the open ocean.[13]

Foraging societies were diverse in their social organization. Hunting-gathering groups were in some cases small and egalitarian, with limited material wealth, and in others more socially stratified as suggested by more elaborate material culture. Evidence at the Niah Caves in Sarawak (in present-day Malaysia on the island of Borneo) shows that the peoples who visited this site may have engaged in a division of labor, based on the variety of hunting and foraging strategies in evidence, including trapping, as well as gathering shellfish and turtles.[14] At the Khok Phanom Di excavation in southwest Thailand, the seminomadic societies that periodically occupied the site left behind prestige objects like axes made of exotic materials and ornaments for elites.[15]

Over time, technologies and the skills to make them spread within and between communities. For example, a unique assemblage of stone tools of characteristic manufacture spread widely around mainland and island Southeast Asia roughly 18,000 years ago.[16] Called the Hoabinhian assemblage of tools (named for Hòa Binh province, located in present-day northern Vietnam), they are made from cobbles (rounded stones, referred to as pebbles) that are flaked on one or both sides.[17] Hoabinhian-style tools found include axes, choppers, mortars, and pestles. Hoabinhian sites have been found as far north as Southern China (near the current border with Myanmar), in Thailand, Laos, Vietnam, and peninsular Malaysia, and on the Indonesian island of Sumatra. That these techniques spread between groups is suggested by the variety of ecological zones in which they are found, from interior forests to coasts and riverbanks. However, it is not known exactly how these techniques spread, whether by imitation, intermarriage, migration, warfare, or demographic expansion.[18]

The Second Wave of Migration: Agricultural Peoples

The second wave of migrants, starting between 2000 and 3000 BCE, were agricultural peoples who brought very different lifestyles and technologies than the paleolithic foragers.[19] Sharp differences exist between the material culture of agricultural populations and that of hunter-gatherers. In areas of long-term, rather than intermittent, settlement, sophisticated forms of pottery-making, evidence of spinning and weaving, and the remains of domesticated plants and animals all point to notably different ways of life. Migrants, probably originating in the Yangtze and Yellow River Valleys, brought agricultural practices to mainland Southeast Asia by traveling south, probably along river valleys.[20] Compelling evidence supports the theory that agriculture arrived in island Southeast Asia via Taiwan, another destination for agricultural migrants from mainland China.[21]

From Taiwan, agricultural peoples made their way to small islands in the Luzon Strait, which are now part of the Philippines, and from there to the big island of Luzon.[22] Evidence including characteristic assemblages of artifacts, DNA, and linguistic relationships suggests that agriculture spread from the Philippines southward into the Malay Archipelago, branching east toward New Guinea and west toward Java and Sumatra. Just as was the case with paleolithic migrants, agricultural populations seem to have been skilled in the use of watercraft. Words associated with ships and navigation trace back to a proto-Austronesian language (most agricultural populations in the region speak Austronesian languages), suggesting the deep integration of maritime activity in the lifeways of the agricultural settlers of Southeast Asia.[23]

Agriculturists brought both techniques for working the land and cultivars new to Southeast Asia. Domesticated rice and foxtail millet originated in the Yangtze and Yellow River areas; both made their way into Southeast Asia. Dryland millet cultivation probably preceded rice agriculture. For example, in Lopburi (in south-central, nonpeninsular Thailand), millet cultivation may have been practiced for 1,000 years before rice agriculture and is still grown in dry highland areas today.[24] The earliest evidence of settled rice agriculture and domesticated animals in Southeast Asia is found along Vietnam's Red and Black Rivers and in Thailand near the Chao Phraya. Evidence of sedentary communities accompanies the spread of rice agriculture throughout areas suitable for its cultivation after roughly 2000 BCE.[25]

Domesticated plants can be usefully understood as technologies in their own right.[26] Plant domestication involves seed selection, a technique for choosing (and ultimately breeding) cultivars with desirable qualities. Therefore, domesticated plants are products of consistent and continuous human intervention, shaped to better meet human needs. With a good nutritional profile and a high degree of genetic variability, rice was ideal for selection. Early populations bred varieties less likely to shatter or drop their seeds when ripe.[27] Over time they developed varieties to suit widely different conditions, from dry rain-fed uplands to irrigated river valleys.

Domesticated plants were among the most environmentally transformative technologies that migrating populations brought with them. Rice agriculture predominated in the river valleys of the mainland and the volcanic soils of the western Malay Archipelago. Although its cultivation and consumption were never universal, it became Southeast Asia's essential staple; the Food and Agriculture Organization of the United Nations shows more than 50 million hectares of land in Southeast Asia planted to rice in 2018.[28]

Hunter-gatherers had mixed responses to the influx of agriculturists. Migrating communities were probably small, and no evidence suggests widespread conflict, although small-scale violence was likely.[29] In many places, including sites in Vietnam and Thailand, hunter-gatherer and agricultural populations seem to have integrated over time.[30] Whether those populations became predominantly agricultural or shifted to a mixed subsistence strategy depended on the ecological circumstances they faced. For example, in places where soils were less fertile and the climate too wet, rice was grown, if at all, via forms of shifting cultivation. Such communities grew rice for no more than a year or two in the same location, allowing an extensive fallow period for the land to regain its fertility. In the eastern Malay Archipelago, cultivators chose more ecologically suitable crops like sago palm (*Metroxylon sagu*) and indigenous fruits and tubers like bananas and yams.[31] On the mainland, rice flourished on the extensive alluvial plains in what is now Thailand, Vietnam, and Myanmar. Yet even here, mixed economies of agriculture, hunting, and foraging were probably the rule.[32] Agricultural ways of life may have provided far more food security than pure foraging, but foraging provided much-needed diversity in local diets.[33]

Although demographic expansion tended to reinforce agricultural ways of life, agriculturalists did not always enjoy a persuasive technical advantage over hunter-gatherers. It is a mistake to imagine the suite of agricultural technologies as an "improvement" over preexisting techniques of subsistence. At the Khok Phanom Di site, populations transitioned to agriculture for a time and then reverted to hunting and gathering sometime before 1500 BCE. The reversion may have been related to the growth of trade with inland agricultural communities.[34] Agricultural migrants were stymied in the uplands of western New Guinea, where existing societies practiced forms of tuber-focused arboriculture and resisted incorporation.[35] Other foraging communities around the region also refused to integrate into settled lifestyles, instead taking refuge in uplands and heavily forested areas that were less appealing to agriculturists.[36]

Foraging ways of life were robust and resilient. Large numbers of foraging communities, including pure hunter-gatherers and others who practiced mixed economies, existed throughout the period covered by this book and up to the present day. Moreover, not all populations who privileged foraging descended from the first wave of migrants. As will be explored further in this chapter, the growth of exchange relationships between foraging and agricultural communities offered advantages that convinced some agricultural communities to shift exclusively to foraging.

Thus, there is no culturally or technologically evolutionary relationship between agricultural and foraging ways of life. It was never a matter of one group possessing skills, knowledge, or technologies superior to others, so much as different sociotechnical opportunities and commitments. As Southeast Asia transitioned from the prehistory to history (understood as a gradual change starting in the first millennium CE), a spectrum of subsistence techniques and technologies emerged, providing a rich technological foundation for life around the region.

Technology and Mobility: Nomadic and Seminomadic Cultures

While the next chapter explores the technologies that supported settled ways of life, for the remainder of this chapter I will delve into the technological cultures of nomadic and seminomadic peoples. The theme of mobility offers a useful starting point to understand technology among nomadic peoples. Mobility highlights how cultures of movement (or lack thereof) are co-constructed with technology, creating characteristic itineraries that define local understandings of space and territory.[37] Transportation technologies like boats or roads are included, but so are other technologies that facilitated or motivated nomadic ways of life and shaped the itineraries of foraging people as they moved across the land- and seascapes of the region. In every case, mobility and technologies that facilitated movement are equally shaped by the opportunities and constraints presented by the natural world.

Understanding the history of technology in nomadic and seminomadic communities is methodologically challenging. The diverse historical trajectories of the many foraging communities make it impossible to generalize or offer a meaningfully "representative" set of experiences or technologies.[38] Evidence may be fragmentary, requiring a synthesis of archaeological, ethnographic, and linguistic data. Scholars may need to consult both written accounts and oral traditions. Archaeological and linguistic data can sometimes offer reliable chronology, as can some written sources and oral traditions. Yet to reliably trace the antiquity of practices or characteristic technologies requires that these elements be either materially durable or embedded in social memory. Modern ethnographies of artisanal, hunting, and foraging practices in contemporary societies offer insights that may be impossible to find in archaeological, written, or oral sources, yet pinning down origins or establishing change over time may be difficult to do. Contemporary communities may also give us only a partial picture. Ancient hunting-gathering societies may have varied in organization, behavior, and culture from those that remain today because they were more widespread in earlier times.[39] Accepting these limitations, the history of Southeast Asia's foraging peoples is nevertheless essential to understanding the broader dynamics of technological change around the region.[40] Therefore, the brief histories discussed in this chapter aim to illustrate aspects of the relationship between mobility and technology in nomadic societies rather than exhaustively describe it.

Anthropologists have long explored the characteristic tool kits that foraging societies use to hunt and collect wild plants and other items necessary for life. These are everyday technologies—basic, frequently used tools needed for subsistence, enjoyment, or ritual. Most nomadic peoples made everyday technologies using readily available materials. Archaeologists can date artifacts made of durable materials like stone, bone, or shell. They can trace the development of technologies with characteristic makes and design, offering a picture of change over time and across space. It is more difficult to historically trace items made of nondurable materials, except by making inferences from more recent practices.

A good example comes from the Semang people of peninsular Malaysia. The Semang are Austroasiatic speakers who have long practiced hunting-gathering as their primary mode of subsistence. Often included in the catch-all category "Orang Asli" (which means "original people") with several other ethnic groups in the region, DNA profiles suggest that they descend from the first wave of migrants into Southeast Asia. The contemporary Semang people use bamboo and rattan extensively. Bamboo is useful for making rafts, shelters, sharp tools for hunting or processing foods, snares, and blowpipes. Blowpipes are more valuable than bows and arrows for hunting in dense forests because they can be used to send projectiles across tighter spaces and far up into the canopy (figure 6). Although seemingly straightforward technologies, blowpipes are designed with the user, the prey, and the poison used for the darts

Figure 6. A Dayak man with a blowpipe.
Courtesy Leiden University Digital Image Collection. Shelfmark
KITLV 1401247. CC BY-SA 4.0, https://creativecommons.org
/licenses/by/4.0/.

in mind. Shorter blowpipes are easier to handle but have less range. Longer blowpipes have greater range and power but require more skill. The poison used for darts in Malaysia, derived from the tree species *Antiaris toxicaria*, takes time to act, making them less useful for birds or other animals that can flee quickly. Blowpipes, therefore, are especially useful to hunt various species of monkeys; in contemporary Malaysia, hunters tend to use a blowpipe that is optimal for the parts of the canopy in which monkeys are typically found.[41] As discussed further in chapter 7, they may also serve as weapons and ritual objects. It is possible that blowpipes were used for hunting in parts of Southeast Asia as early as 12,000 years ago.[42]

Rattan (the name for approximately 600 species of climbing palm or *Calamoideae*) is strong and useful for everyday life. Southeast Asian peoples have used rattan to make tools like twine, baskets, mats, hunting snares, and even, when correctly handled, bridges and shelters. Although it is difficult to discern just how old any given use of rattan is, rattan products and drugs made from rattan resin (under the name "dragon's blood") were traded to China no later than 1,000 years ago.[43] It is reasonable to suppose that daily use of rattan probably dates back much earlier.

These tool kits are notable because they facilitated mobility—making hunting and foraging possible. The immaterial technological knowledge they carried was just as important. Making useful objects on the fly allowed nomadic societies to balance portability with time spent crafting objects for hunting, fishing, or collecting.[44] As with many everyday technologies, the seeming simplicity of rapidly constructed technologies should not be taken for granted. Recall, for instance, that relatively simple watercraft made distant sea crossings—no mean feat. There was no tradition of making objects that required more infrastructure, such as metal items or sophisticated ceramics. Metal industries grew only in agricultural communities.[45]

Technologies like boats, bridges, and hunting gear facilitated desired mobility. However, other technologies, especially those important for subsistence and social flourishing, made certain itineraries of movement more desirable. The technique of swiddening is a valuable example. Swiddening is a form of agriculture where crops are planted in a single area for a limited amount of time, often no more than one to two years, followed by some number of years of fallow. Fallowing periods and cultivation cycles vary depending on the crop and environment. For example, in areas of Sulawesi where soil is low in nutrients, farming groups practice long-fallow swiddening, with 15–20 years of fallow after cultivation. After the fallow period, the land is cultivated again, and the cycle continues. Swidden plots are not permanent in the way that sed-

Figure 7. A modern swidden field in Yunnan Province, China. Notice the variety of stages of growth represented from mature forest to newly cleared land.

By Desmanthus4food via Wikimedia Commons. CC BY-SA 3.0, https://creativecommons.org/licenses/by/3.0/.

entary agriculture is, although they may be used repeatedly by the same people for generations (figure 7).[46] Practiced skillfully, swiddening is often the most sustainable form of agriculture for areas of marginal fertility such as rain forests and uplands where soils are fragile. Exhausted soil can lead to permanent loss; tropical forests cannot reclaim overworked land.[47] Offering food security at the modest cost of increased community mobility, swiddening makes seminomadic lifestyles in fragile environments sustainable and desirable.

It can be difficult to ascertain how long any given group has included swiddening in their subsistence strategies. Swiddening in some areas may date back to the Neolithic period, as farmers expanded from the most fertile rice areas into more marginal areas.[48] Yet not all swiddening is ancient. Among those peoples of Borneo called the Punan, a catch-all category for groups of Austronesian-speaking Bornean nomadic hunter-gatherers, swiddening has probably been part of their subsistence strategies only for the past two centuries.[49] In Timor-Leste, historians from the early to mid-twentieth century tended to assume that the development of agriculture was a fully linear process leading from "primitive" foraging or swiddening to "advanced" sedentary agriculture. In fact, it was likely adopted during the early twentieth century as a response to colonial rule.[50] Swiddening is far from primitive, requiring a deep understanding of soils and ecologies and sophisticated social relationships to enforce fallowing. Rather than embrace misleading assumptions about linear technological change, it is more helpful to think of various approaches to agriculture as strategies that (assuming knowledge is maintained across generations) can be taken up to fit a variety of political circumstances.[51]

Technologies that supported or motivated mobility among nomadic and seminomadic peoples may have originated as part of local subsistence strategies. But they later served as a foundation for trade with sedentary communities. Eventually, forest and sea products formed the foundation of long-distance trade, particularly with Chinese merchants active in island and mainland Southeast Asia no later than the Qin dynasty (221–206 BCE). They sought materials for medicine and perfume, including forest resins, the scented wood called gharu or agar wood (found in the Malay Archipelago), tortoiseshell, trepang (a variety of sea cucumber), pearls, and coral.[52] Sedentary communities served as intermediaries between collectors and foreign merchants, as was the case in the Tanjay River area of the central Philippines. There, upland foraging peoples entered into political and economic relationships with coastal peoples, providing forest products and game in exchange for grain and prestige goods like Chinese porcelains and jewelry.[53] Such relationships became the foundation of a valuable trade that only increased throughout the first millennium CE.

Mutualistic relationships reinforced the coexistence of mobile and sedentary ways of life and technological specialization even as it subtly modified how each lived. Technological specialization might help to reinforce the boundaries between communities. For example, modern Semang peoples, Austroasiatic-speaking hunter-gatherers who live in peninsular Malaysia, participate in economically unequal trade to prevent outsiders from moving onto their land. For the Semang, their knowledge and skills in foraging are not just economically valuable; they are essential tools for protecting the lands they move through and their chosen way of life.[54] Foraging techniques, tools, and knowledge were (and are) no mere remnant of a receding past but dynamic technologies whose use shaped environments and the emergence of Southeast Asia as a site for long-distance trade.

Conclusion

From prehistory to the early years of the first millennium, human settlement created a mosaic of farming and foraging populations in Southeast Asia. Despite the demographic expansion of agriculturists, foraging communities, including hunter-gatherers, did not vanish from the landscape. The interactions between foragers and farmers and the rise of significant foreign trade in products only foraging societies could reliably provide strongly informed the patterns of circulation that emerged around the region. Moreover, the interdependence between these different communities meant that sociotechnical change or shifting economic conditions from either community could reverberate with the other. The fullest picture of technology in Southeast Asian history requires close attention to the material infrastructures of both nomadic and settled communities and the technologically mediated engagements that arose over time.

CHAPTER 2

Agriculture and Trade

Technological Change and Emerging Urban Centers

S EDENTARY COMMUNITIES dramatically transformed human life and wider environments across Southeast Asia. Settled populations modified their environments in ways completely different from the foraging communities that preceded them. They developed intensive forms of agriculture, redirected flows of water, and produced innovations that made long-term habitation possible and desirable. They produced a plethora of material objects that were practical, prestigious, or both, using techniques in areas like metalworking, weaving, pottery-making, brick-making, and more. Densely populated urban centers would ultimately become centers of trade and Southeast Asia's first states, creating new patterns of mobility and circulation that could complement or challenge the mobility and ways of life of preexisting foraging communities. This chapter explores the coproduction of urban centers with the technologies that served the vital purposes of habitation, provisioning, and social cohesion.

Apart from the insight they provide into developing technological ways of life, paying attention to growth of urban centers draws attention to the emerging assemblages of technology and people that enable connections across distances and boundaries of language and culture. Science and technology studies (STS) scholars have long explored the ways that agency is distributed across networks of people and things.[1] The idea of assemblage nicely captures the ongoing, sometimes ephemeral and sometimes durable arrangements of people and things that make possible new kinds of relationships and circulations—social, political, commercial, or cultural. Over time, these patterns of circulations and relationships, emerging from assemblages of people and technologies, constitute the region.

Technology and Emerging Urban Centers

Although agriculture was established in parts of Southeast Asia in Neolithic times, urban centers started to emerge between 300 and 500 CE. Acting as focal points for both trade and ritual, urban settlements were sites of technological innovation, including the development of new artifacts and techniques as well as ways of organizing labor. Yet no single sociotechnical blueprint dominated.[2] Instead, multiple environmental, cultural, and economic factors influenced how societies assembled diverse technologies into interconnected infrastructures to support urban life. Earlier scholarship emphasized material and cultural influences from the Indic societies of South Asia in the emergence of these cities, and indeed those influences mattered.[3] But more recent archaeological study has made the role of local actors clearer, and therefore, the place of local innovation in urbanization.

One schema for distinguishing urban settlements from villages is: "evidence of a considerable economic and social complexity, e.g. the existence of specialised craft groups and non-food producers, made possible by food security and surplus, either locally sourced or from elsewhere; the emergence of socioeconomic inequalities, leading to hierarchical relationships and evidence of large-scale coordinated activities (such as the construction of monuments, walls and irrigation works); and the urbanising community as a central place within its area."[4] Modes of providing food and water and the development of craft specialties offer insight into the varieties of sociotechnical arrangements that supported Southeast Asian urbanization.

Provisioning: Food and Water

Reliable food surpluses and sufficient water were essential for the flourishing of sedentary populations in general. Still, techniques and technologies for providing these fundamentals of subsistence changed with the emergence of denser urban centers. Through much of mainland and parts of island Southeast Asia, rice agriculture became central for the basic subsistence of such populations; it offers a valuable lens through which to explore the ways that diverse agricultural techniques and technologies were coproduced with urban societies.

Neolithic farmers probably grew cultivars of domesticated rice in swamps or other natural wetlands similar to sites where the wild rice domesticated by the earliest Asian farmers evolved. Over time, farmers took advantage of the inherent genetic mutability of rice to develop cultivars viable in a wider variety of environments, including those areas where the only water available comes from unpredictable rainfall.[5] However, most dense urban populations were supported by wet-rice cultivation, the practice of growing rice on flooded fields. Wet rice cultivation techniques require careful preparation of the soil substrate to hold water while plants are young. Water supports plant growth and drowns undesirable weeds and pests as the crops grow. When the rice ripens toward harvest time, the fields need to dry naturally or be drained. Wet-rice cultivation offers significantly higher yields than other methods.[6]

Diverse technologies or techniques were used around the region to provide or drain water from the fields, depending to a great extent on the environmental circumstances. Those who lived on natural floodplains took advantage of the predictable flooding patterns, as farmers would plant rice either in anticipation of or following the flood.[7] Bunded fields (those with built-up perimeters) would contain the floodwaters to the proper depth. Some created tanks or ponds to collect water until it was needed. Like the seed selection techniques mentioned earlier, farmers' knowledge of both weather and the response of rice plants to water across their life cycle is evident in their development of inundation techniques.

Figure 8. A terraced paddy field in Bali.
By V. Epiney via Wikimedia Commons. CC BY-SA 2.0, https://creativecommons.org/licenses/by/2.0/.

It is no accident that many of Southeast Asia's earliest urban centers emerged near the floodplains of major rivers. Evidence suggests that rice farmers worked the floodplain areas of the Chao Phraya River (the largest river in present-day Thailand) as early as 200 CE and used the bunded field technique regularly by no later than 700 CE (figure 8).[8]

Between 200 and 1000 CE, the Tibeto-Burman-speaking Pyu were among the earliest people to practice flooded rice agriculture in the central area of what is now the nation of Myanmar. Using the floodwaters of the Ayeyarwady and two of its major tributaries, the Mu and Chindwin Rivers, Pyu agriculture provided food surpluses to support the sophisticated walled cities that flourished during this era.[9] They significantly influenced later Burmese states artistically, politically, religiously, and technologically.[10] Likewise, in the region now occupied by southern Vietnam and Cambodia, farmers in the Kingdom of Champa (192–1832 CE) took advantage of predictable seasonal flooding along the Mekong and Tonle Sap (a lake fed by a seasonally flooding river) for rice cultivation. Floodplain farming in this area included the use of bunded fields, irrigation tanks to store water out of season, and careful timing to either anticipate or follow the flood.[11] Inundated wet rice cultivation spread to island Southeast Asia as well, probably by way of the Philippines, via the Bicol area of southern Luzon.[12]

In places farther from areas of natural flooding or in places where such flooding was too severe, some Southeast Asian societies supported rice agriculture through the installation of permanent irrigation systems. In Myanmar, Vietnam, Thailand, and Java, societies built and maintained canals and catchment systems to divert water from major rivers or mountain streams. Permanent works required considerably more upkeep and labor to support. But reliable water management made those communities less vulnerable to both drought and flood. In some areas, permanent irrigation works facilitated two rice crops each year, which greatly benefited those seeking larger food surpluses.[13]

What drove these technological choices? Although intensive modes of agriculture could support denser populations, population pressure appears not to have been a major factor much before the nineteenth century. Environmental challenges seem to have played a larger role, albeit one affected by population increases or decreases. Different forms of rice agriculture were practiced in the same areas by the same people, depending on conditions, including weather.[14] Even in some urban areas, less labor-intensive dryland farming could be a viable option, especially when other food sources were readily available. The coastal city of Kedah on the Malay Peninsula provides a useful example. In the years around 500 CE, Kedah's thriving trade-based economy was supported by dryland rice cultivation in the hills around the region. Relying on their relationships with autonomous networks of forest collectors to supply the products they traded, the people of Kedah found dryland rice sufficient to provide food security for their cosmopolitan city. Only when the resident population started to grow beyond the hill regions' capacity to supply them and the dryland areas began to experience more severe erosion and soil depletion did they turn to irrigation. Irrigated rice became the norm with the formation of a floodplain in the region after 1200 CE.[15]

By contrast, the Pyu (mentioned previously), c. 500 CE, used extensive

Figure 9. The remains of a gate of the Pyu city of Sri Ksetra in Myanmar. Brick construction of city walls and canals were linked.
By Jakub Hałun via Wikimedia Commons. CC BY-SA 4.0, https://creativecommons.org/licenses/by /4.0/.

works to divert water from rivers and capture rain in tanks, weirs, moats, and ponds for irrigation, protection, and drinking water. Importantly, these systems reduced erosion in agricultural areas, a significant local challenge. The Pyu works were so well suited to their environment and so functional that a few are still in use today.[16] Other technological capacities within Pyu society may also have influenced the ways they envisioned possibilities for water storage. Their water infrastructure clearly drew on similar technologies as those already used in buildings and city walls, including an innovative use of bricks to create more durable linings for moats. The emergence of better walls and better waterworks were linked; walls constructed around cities and burial terraces were physically connected to canals and moats (figures 9 and 10).[17]

Growing social complexity and stratification common to emergent urban communities often (but not always) correlated with the establishment of permanent irrigation works. For example, in the town of

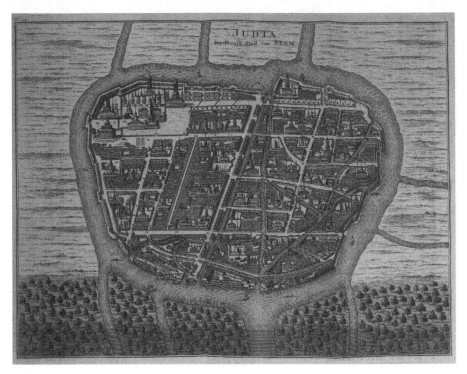

Figure 10. This European depiction of the city of Ayudhya from 1726 (titled "Judia, De Hoofd Stad of Siam," "Ayudhya the Capital City of Siam") shows some of the characteristic features of Southeast Asian urban centers, including a moat around the city, protective brick walls, and canals.
In François Valentijn, *Oud En Nieuw Oost-Indien*, v. 3, pt. 2. Public domain.

Phum Lovea (in present-day northwestern Cambodia), the appearance in 500 CE of improved bunded rice fields served by permanent irrigation works was associated with the rise of elites and a more centralized organization of society.[18] Did growing social stratification and centralized governance provide the resources to erect permanent irrigation systems? Or did the increased yields associated with irrigated lands stimulate sociotechnical change? Irrigation works may have been an opportunistic add-on to water supplies intended to protect cities. Or food surpluses from effective irrigation systems may have made settlements more resilient to irregular weather or other problems, allowing for consistent growth and security among populations, both of which could stimulate social complexity. It is not possible to say with certainty, but it is reasonable to assume sophisticated irrigation infrastructure was coproduced with stratified forms of social organization (figure 10).

As the case of Kedah shows, it was never necessary to build sophisticated irrigation works to develop socially complex cities, but in the right environmental context, they proved to be sufficiently helpful and versatile to be worth the investment.

Artisanal Developments

Food surpluses provided the wherewithal to support dedicated craft specialists, another area of considerable technological dynamism. Artisans produced goods both for local use and very often for trade. For example, in the Mun Valley of Northeast Thailand, salt has been produced on a commercial scale since at least 300 CE. Salt was likely made by collecting topsoil, filtering water through it, and boiling off the water.[19] Surplus salt was used to salt fish, a traded commodity that provided a readily portable supply of protein and facilitated long-distance mobility around the Southeast Asian mainland.[20] Iron smithing and metalworking also developed at the cusp of the historic period, possibly influenced by South Asian techniques, with nails, jewelry, and other metal objects circulating widely.[21]

A celebrated example of the linkage between trade and craft specialty is the sophisticated bronze and iron working of the Đông Sơn culture that flourished between 1000 BCE and 1 BCE in northern Vietnam's Red River Valley area. Named for a village where evidence of their bronze manufactures was first found, the Đông Sơn peoples were a socially stratified society whose economy was based on irrigated rice agriculture via purpose-built canals.[22] They left behind beautifully crafted bronze items, including plowshares, knives, and other tools, bowls, and bracelets. The most famous artifacts are their remarkable bronze drums (figure 11).

Made using a lost-wax casting method, these bronze drums were created for ritual purposes (and are often found in burials), such as the

Figure 11. Close-up of a Đông Sơn drum, also known as a Heger 1 drum.
By VuThiAnh via Wikimedia Commons. Public domain.

investment of kings. Drums from the early period are decorated with images of daily life, such as people on horseback, women weavers, houses raised on piles, rice harvests, and diverse animals. There are depictions of warfare, including canoes outfitted for river-based fighting. Even the drums themselves appear in portrayals of rituals.[23] Some drums made when the Chinese ruled the area include figures in recognizable Han-style clothing.[24] Although valued for ritual, they were also made for trade, sometimes even made to order.[25] Drums have been found across mainland Southeast Asia, in South China, and even in the Malay Archipelago, especially around the Straits of Malacca.[26] Food surpluses and the security of urban settlements offered the foundation on which such sophisticated craft techniques could emerge. Seemingly different areas of innovation such as food, water supply, and artisanal goods were interconnected, coproduced with urbanizing ways of life.

The Infrastructures of Trade

Trade, like urban settlement, motivated and was motivated by technological change. Trade was a significant factor in Southeast Asian life deep into prehistory. For example, archaeologists have found marine shell ornaments more than 1,000 km from any coast and other rare materials at great distances from the closest possible point of origin. The development of valuable items and the construction of infrastructures to support trade involved notable technological investments. The resulting assemblages of technology and people reinforced and expanded commercial relationships.

Maritime technologies like boats and techniques like navigation provided the foundation for most long-distance trade in Southeast Asia. Although little material evidence remains to ascertain specific technologies in use, other archaeological findings demonstrate that Southeast Asians made ocean crossings at very early dates. Southeast Asian navigators had neither compasses nor written charts. They may have employed the kinds of methods that Austronesian navigators in the Pacific islands developed: observations of land, sky, and sea, and the ability to interpret these observations reliably, passed down via oral transmission.[27] Technologies facilitating long-distance trade around the Asian mainland developed around the possibilities and restrictions of the monsoon.[28] Sailing in ships that could not easily travel against prevailing winds, by 1000 BCE, traders had learned to predict the seasonal monsoons (blowing easterly in the winter and westerly in the summer). Simple infrastructures like warehouses made it easier to ride out the unfavorable monsoon in a foreign port, gathering goods for the return voyage. Eventually, Southeast Asian traders sailed as far west as Madagascar and routinely to Bengal.[29]

Ships capable of sailing long distances with adequate cargo were the infrastructural backbone of expanding long-distance trade. In Southeast Asian seas, the *jong* played this role (figure 12). These two- or four-masted wooden ships were constructed using wooden dowels and rattan sails.[30] Evidence suggests that by the third century CE, they could carry at least 250 metric tons of cargo; by 1500 CE, a capacity of 400–500 met-

Figure 12. Jong-style ships in the Harbor at Batavia, c. 1726, detail.
In François Valentijn, *Oud En Nieuw Oost-Indien*, v. 4, pt. 2. Public domain.

ric tons was common, with some sources reporting ship capacities as large as 1,000 metric tons.[31] In the sixteenth century, Portuguese observers noted that their guns could not puncture the multiple layers of sheathing commonly comprising these ships' hulls.[32] Notably, the *jong* normally had a dual rudder rather than the single rudder commonly seen in Chinese ships. While one of the most famous representations of a ship from the period, the relief carved into the Buddhist monument of Borobudur in central Java, c. 800 CE, shows outriggers, outriggers were probably only used in ships sailing calmer, protected waters, unlike the larger workhorses of the India and China trade.[33]

Did the design of these ships originate in Southeast Asia, or did models from elsewhere influence them? Based on the wooden joinery, the unique steering mechanism, and chronological and linguistic evidence, there is good reason to believe that Southeast Asian ships were indigenous to the region and not derived from Chinese junks.[34] Linguistic evidence shows that Indian peoples adopted Austronesian nautical terms rather than vice versa, suggesting that Southeast Asians were recognized as skilled seafarers from an early period.[35] We have little evidence about where shipbuilding took place during the first millennium CE. However, there is later evidence of major shipbuilding centers on the northeast coast of Java (near present-day Cirebon), the southern coast of Borneo, and in Bago (also transliterated as Pegu), on the coast of the present-day state of Myanmar.[36]

Growing maritime trade spurred the development of more technologically sophisticated cities, demonstrating the interconnections that were emerging between seemingly distinct areas of technological innovation. Consider the example of the state that Chinese traders called Funan, one of the earliest major coastal polities that emerged in the region. (The indigenous name has been lost.)[37] Around 50 CE, Funan was established as a trading center near the Mekong Delta in southern Vietnam. Its frequent mention in Chinese and Cham records and Chinese, South Asian, and Roman artifacts found in the area confirm its cosmopolitan character and significance across Southeast Asia.[38] Chinese traders turned to Funan for trade with South Asia when warfare in western China made the overland route too dangerous. More than transshipping South Asian products, Funan facilitated the entry of Southeast Asian forest products into Chinese and Indian markets, including aromatic resins, sandalwood, camphor, and gold objects, as well as later-cultivated products like cloves.[39] Chinese merchants sought out forest products as early as 200 BCE for use in medicine and perfumes; Funan brought them to wider markets, increasing demand for items largely collected by nomadic and seminomadic peoples but shipped via settled coastal communities.[40]

Funan's ability to prosper as a trading city relied on infrastructural technologies, especially those that supported the creation of food surpluses adequate for populations of traders and their crews.[41] The exact picture of agriculture in Funan is unclear, but farmers likely took advantage of seasonal flooding to practice wet-rice agriculture using broadcasted seed. Irrigation systems and long-term water storage probably became common in the region by the fifth or sixth century CE.[42] Imported bronze plowshares from the Đông Sơn region may have helped increase yields. The Funanese opened land for agricultural production

Figure 13. Archaeological site located in Óc Eo, An Giang Province of southern Vietnam. Located in the Mekong Delta, Óc Eo was a port in the kingdom of Funan. By Bùi Thụy Đào Nguyên via Wikimedia Commons. CC BY-SA 3.0, https://creativecommons.org /licenses/by/3.0/.

using hydraulic technologies to drain saltwater areas near the coast.[43] Simultaneously, they built canals whose main purpose was transport between the main trading sites in the larger kingdom, which may also have drained land to make it viable for agriculture.[44] Funan's leaders constructed palisaded cities, developed harbors, and employed up-to-date brick construction for large temples and monuments (figure 13). Although requiring considerable labor, the techniques and methods they used were not new. Instead, they implemented on a larger scale technologies similar to those in use in smaller urban settlements.

In addition to motivating the spread of characteristic forms of infrastructure, trade also stimulated the production of new commodities, a process that may have required that artisans learn or invent new techniques or technologies. Funan's artisans became adept at creating "imitations" of exotic trade goods. When Funanese artisans began to create high-quality counterfeits of desirable Indian goods such as glass beads and gold, tin, and bronze ornaments, traders happily began dealing in them.[45] While no evidence allows us to trace precisely how the process of artisanal learning occurred (i.e., through the migration of skilled manufacturers or local innovations based on craft knowledge and experimentation), it nevertheless points to a technologically dynamic environment. Funanese goods appear to have been traded across long distances.[46]

By the fourth century CE, other coastal trading centers farther south were emerging to compete with Funan. Supported by intensifying and more extensive agriculture in their hinterlands and improving maritime technologies, newer trading kingdoms eventually eclipsed Funan's power. One of the most significant of these was Srivijaya, whose capital was located at the site of the present-day city of Palembang on the southeast coast of the island of Sumatra. Relying on both agriculturists from the rich hinterland of the Musi River and forest collectors in the Sumatran interior, Srivijaya gained wealth and influence in the middle of the seventh century and flourished for the next five hundred years.[47]

The growth of such cities, co-emergent with a significant expansion of trade between 900 and 1300 CE, triggered a massive social, political, and economic transformation in Southeast Asian societies, an early "age of commerce."[48] The changing economy of China drove much of this expansion. The Song dynasty (960–1279) taxed oceangoing merchant ships and sponsored numerous trade missions to Southeast Asian kingdoms, including Srivijaya and Champa. China exchanged their unparalleled ceramics and lacquer goods, silk, rame (fibers from the nettle family), and cotton for Southeast Asian forest and sea products, rice, pepper, and sugar.[49] By the 1200s, Chinese migrants (fleeing the fall of the Song dynasty and the rise of the Yuan) set up significant settlements on the Malay Peninsula and Champa, reinforcing Chinese trade. The Fatimid Caliphate (909–1171) in North Africa expanded into the region at the same time.[50] Arab and Persian traders became particularly visible in the Cham court and Srivijaya.[51]

This expansion spurred infrastructural development in coastal areas. Some capital cities even moved closer to the coasts. For example, the Mataram Kingdom (716–1016 CE) on Java expanded into the Brantas delta in the northeastern part of the island after experiencing significant economic booms in the ninth century.[52] Likewise, in the early twelfth century, the kingdom of Đại Việt (in present-day northern Vietnam) established a new port at Vân Đôn, located on the Gulf of Tonkin, where they regularly traded with Chinese, Cham, and Muslim traders.[53]

Technological change in artisanal fields followed the expansion of trade. Changes in local ceramics manufacture around Southeast Asia demonstrate this point. In the coastal regions of Đại Việt, and possibly also in Angkor, migrant Chinese craft specialists directly influenced local ceramics manufacture by doing business with local ceramics makers. Locals modified their techniques to create products closer to the coveted Chinese types. On Java, although there is no evidence of Chinese migration, the demonstration effect of fine Chinese ceramics is nevertheless clear. Not only did Javanese ceramics artisans imitate the characteristic shape and design of Chinese goods, but they also adopted Chinese tools and techniques such as the potter's wheel over older paddle and anvil techniques. How the potter's wheel made its way to Java (whether in the hands of experts or as a trade good) is unclear.[54] Likewise, Javanese weavers imported cotton and indigo from South Asia by the ninth century CE and had modified their looms and adapted Indian patterns to imitate highly regarded Indian cloth.[55] This combination of circulating knowledge in the hands of migrant or itinerant craftworkers and the demonstration effect provided by appealing foreign goods drove the development of artisanal knowledge and techniques.

The opportunities offered by trade stimulated agricultural change as well, including the expansion of preexisting practices to new lands and growing new crops for developing export markets. Temple networks may have played a significant logistical role in spreading agricultural practices to new areas. Đại Việt's expansion toward the coast was made possible by the intensive development of agriculture in Buddhist temple networks. The Lý kings, who reigned between 1009 and 1072, offered land to build as many as 1,000 new Buddhist temples, probably as a way to integrate disparate lands under their rule. Although little research exists on the economic role of these temples in Đại Việt, they may have served a similar purpose as those that arose in Bagan (a historical kingdom located in what is now Myanmar).[56] Buddhist temples were more than ritual sites. They distributed land and seeds to farmers and built infrastructure like irrigation works to support agriculture. Monastic centers attracted artisans to support the varying needs of both laypeople and religious communities. The growing agricultural base of the northern interior created demand for fine Chinese products, reinforcing trade.[57]

A similar expansion of agriculture around temple networks also occurred in the Javanese Kingdom of Mataram (716–1016). Temple complexes encouraged the growth of agriculture in the region in the 700s and 800s. By the 900s, when the kingdom was more oriented toward external trade, farmers began double-cropping rice to supply export markets. They also expanded into desirable trade crops like black pepper (*Piper nigrum*), safflower (*Carthamus tinctorius*, used for dyes), and sugar.[58] Little evidence is left to explain exactly how farmers obtained and learned to grow these new crops, but their significance for Mataram's agricultural economy is indisputable.

As political and mercantile power moved around Southeast Asia from roughly 500 to 1500 CE, a similar technological recipe for success is evident. It included deploying techniques and technologies proven to intensify or expand agriculture, building infrastructures to facilitate or draw trade, innovating craft techniques to produce goods desirable for distant markets, and performing the technical work required to expand forest collecting. Even as much of this technological dynamism was bottom-up, driven by disparate individuals and communities, the growing circulation of people, things, and ideas resulted in broadly recognizable similarities and many local variations on an interconnected technological culture.

Infrastructure and State Power

Significant social and economic investment in infrastructure did more than support trade; it also underpinned the emergence of the "charter states" of Southeast Asia between 900 and 1350 CE.[59] The term "charter state" indicates how these polities created durable religious, political, and administrative models taken up by subsequent kingdoms. States such as Bagan (Myanmar), Đại Việt (Vietnam), Angkor (Cambodia), and Srivijaya (Indonesia) have rated attention from scholars in part because of their scale: their geographic, political, commercial, and cultural reach was remarkably large, outstripping that of their predecessors. Investments in infrastructure helped support the large scale of these polities.

Consider, for example, the Buddhist Kingdom of Bagan (849–1287), whose geographical reach was close to the boundaries of the present-day nation of Myanmar. They constructed more than 10,000 religious monuments, including temples, monasteries, and stupas, demonstrating the state's spiritual authority and providing a material foundation for shared culture across a disparate region (figures 14, 15, and 16). The food surpluses needed to support the densely populated capital region were created with the help of sophisticated irrigation works.[60] Likewise, the leaders of the Khmer Empire (802–1351 CE) engaged in massive public works projects in the city of Angkor, including the construction of two huge *barays* (reservoirs) for storing water and more than 1,000 km of canals for transportation around the empire. Such infrastructure underpinned the kingdom's growth to over 3,000 km^2 by the thirteenth century.[61]

These kingdoms were larger and could exercise greater control over their populations than did earlier states. Local innovation (following this general recipe) could enhance both the pragmatic and ritual capacities

Figure 14. The Temple of Borobudur in Central Java. It is the largest Buddhist temple in the world and was built in the ninth century CE.
By Gunawan Kartapranata via Wikimedia Commons. CC BY-SA 3.0, https://creativecommons.org/licenses/by-sa/3.0/.

Figure 15. A view of multiple temples on the north plain of Bagan.
By Adam Jones from Kelowna, BC, Canada, via Wikimedia Commons. CC BY-SA 2.0, https://
creativecommons.org/licenses/by-sa/2.0/.

of these infrastructures. However, infrastructure also creates vulnerabilities. Exploring how different kingdoms innovated to make and maintain infrastructure demonstrates both the affordances and fragilities of the charter states' sociotechnical ways of life.[62]

Temples

Temples, as suggested earlier, served spiritual, political, and practical ends. Temples encouraged cultural identification with the spiritual/ state authority that was meaningful even at some distance from the capital. Pragmatically, they sometimes served as outposts for agricultural intensification, increasing food supplies for the state. They required significant labor (by enslaved people and through systems of corvée) and materials, and careful structural and ritual design. They became an

Figure 16. Angkor Wat, the central temple complex of the capital city of the Khmer Empire.
By Bjørn Christian Tørrissen via Wikimedia Commons. CC BY-SA 4.0, https://creativecommons.org/licenses/by-sa/4.0/.

important site where local and foreign influences hybridized into forms characteristic of Southeast Asian culture.

The design and cosmological significance of temple complexes in the charter states may have derived from Indian Hindu and Buddhist precedents, but they showed important local innovations. For instance, in Bagan, Burmese temples employed certain fundamental features of Indian architecture, like the *gu* (the Burmese word for "cave"), a hollow interior area that houses the Buddha's image. However, Burmese builders sought to increase the size of the *gu*, experimenting with the use of the arch (which South Asians did not use) and the vault. The Burmese may have drawn on Central Asian or Chinese antecedents or even earlier Pyu practices to make their impressive open spaces.[63] The focus on technical construction challenges like arches and vaults suggests how important highly skilled builders were for the Burmese state.[64]

Another construction technique used to enhance the ritual value of temples was the stepped pyramid, as seen in the Hindu temple of Angkor Wat (the ritual center of the city of Angkor) and the Buddhist temple of Borobudur built by the Sailendra kings on Java. This feature probably symbolized Mount Meru, the center of the world in Hindu and Buddhist cosmology. As with the Burmese gu, the stepped pyramids of Southeast Asia showed considerable elaboration compared with South Asian precedents.[65] Angkor's stepped pyramids may have been influenced more by Borobudur than by South Asian temples, suggesting the growing mutual influence of Southeast Asian kingdoms.[66]

Large kingdoms often supported infrastructure designed to support urban centers and interconnect distant locales, creating new engineering challenges. Although infrastructures like water storage and canals drew on well-understood techniques, these infrastructures aimed to support much larger populations. Like the barays that stored water for Angkor, some technologies were physically larger than predecessor technology, which required careful engineering. Other technologies, like the canals of the Khmer Empire, were linked together in long-distance networks of unprecedented size and complexity, possibly encouraging economic and technological interdependence.[67] Such networked infrastructures provided access and mobility but were simultaneously a point of vulnerability if maintenance failed.

How were such networks managed? Although central authorities did control some elements of these networks, few were entirely managed from the top down. Arguments claiming that Asian empires became "despotic" in order to build and support the large, centrally controlled hydraulic works have long been disproved by the evidence.[68] The construction, design, and use of these hydraulic technologies point to more complex arrangements.

That interconnected technological systems did not have to be centrally organized in every respect accords well with the leadership practices in Southeast Asia in this era. Many states were organized into a "mandala system." Rulers exercised direct control over relatively small areas with decreasing levels of control further away from the center of power. Certain areas might be "ruled" by multiple kingdoms with varying degrees of intensive involvement and demands.[69] Charter states and other polities in Southeast Asia crafted various models for managing water on a large scale in keeping with these political models. The following cases explore the varying strategies that societies around the region used to build and maintain hydraulic systems on a scale larger than a city or village.

Hydraulic Technologies: Java and Bali

On the island of Java, rice cultivation predated the emergence of major kingdoms like the Kingdom of Mataram (discussed earlier) or the later Majapahit (1293–1517). Irrigation at the village level, or sometimes cooperatively with other villages, commonly used bamboo, logs, or piles of stone for works designed to be temporary. On Java, swift-flowing seasonal streams and occasional volcanic eruptions could disrupt even robustly constructed irrigation works. Temporary, rather than permanent, works made sense, and local labor and expertise maintained them.[70]

However, by 800 CE, more sophisticated permanent waterworks, including dams, tunnels, and canals, were built in areas close to centers of power. Authorities would call on farmers' labor, which was owed as a form of tax, to build and maintain these works. These systems were designed for irrigation and protection in times of war; flooding could be triggered in defense of the *kraton*, or palace.[71] Little is known of the expertise of royal overseers of these projects, but it appears that the engineering experts were not residents of the palace-city but instead villagers. Given Java's generally sparse population, authorities had to be careful not to overtax local populations. In some cases, peasants who were asked to maintain important dams over the long term were freed from all other tax obligations via land grants.[72] Except for military use, village authorities took charge of water distribution. On Java, therefore, both centralized and distributed management of water coexisted.[73]

On the neighboring island of Bali, highly sophisticated institutions of intervillage cooperation called *subaks* arose in the ninth century. Authority for making decisions about sharing water and building irrigation works was entirely distributed—there was no central authority involved at all. Intervillage councils jointly determined how to distribute water that flowed from high-altitude lakes and mobilized workers to build and maintain infrastructure without any hierarchical central authority.[74] The collective operation of subaks maximized yields for all involved. When rice planting was timed properly across multiple villages, water was used efficiently, and pests failed to gain a foothold thanks to the coordinated flooding of fields. Although subaks were as subject to the vagaries of human conflict as any other institution, functional subaks tended to make societies prosperous. The subak water management system produced some of the highest rice yields in the premodern world and is still in use today, making it a particularly durable and resilient form of water management.[75]

Hydraulic Technologies: Angkor

By contrast with Bali, centralized development played a significant role in making the city of Angkor a flourishing center for culture and trade. Kingly deployment of corvée and enslaved labor built the large barays, the massive temple complex at Angkor Wat, and parts of the canal system.[76] Yet even here, water infrastructure operating farther from the capital emerged in a more piecemeal and autonomous manner, driven by local needs and locally recruited labor. By 800 CE, when Jayavarman II founded the kingdom, people in the region around Angkor had long practiced rice agriculture in the floodplain of the Tonle Sap.[77] In the city, elites oversaw the maintenance of barays and canals.[78] Moving away from the center, canals, tanks, and irrigation ponds were in all likelihood built and operated locally by the people who used them. In some cases, farmers may also have constructed and maintained the modest local temples that formed the center of each community.[79]

Could Angkor's impressive infrastructure have contributed to its ultimate decline? There is some reason to think this might be so, although scholars have intensely debated the reasons behind the Khmer Empire's collapse. Traditional stories emphasize the devastating invasion in 1431 by the kingdom of Ayudhya (1350–1767), a Tai kingdom that stretched across much of the area of modern-day Thailand. Environmental damage, including the relentless clearing of forested land to expand agriculture, subsequent erosion, and topsoil degradation, might have harmed rice yields necessary for the city's support.[80] Yet studies of Angkor's hydraulic infrastructure reveal significant vulnerability to chronic urban stresses and climatic shifts.[81] Evidence shows that the hydraulic infrastructure was already breaking down almost 100 years before the Ayudhyan invasion. Angkor's demise may have been protracted; land use in the city points to the gradual abandonment of Angkor by its urban elites over a hundred or more years. The decline of Angkor's infrastructure may have resulted from this long-term decline in elite populations and the slow collapse of its centralized systems of infrastructure maintenance.[82]

Hydraulic Technologies: Bagan

In the kingdom of Bagan (849–1287 CE), hydraulic technologies supported the all-important work of agricultural intensification and expansion, which was far more important than long-distance maritime trade.[83] Here, too, despite the importance of infrastructures, the process for building and maintaining them was not purely centralized. As was also true in parts of Vietnam, private donations of large tracts of land to Buddhist temples were key to developing a thriving and expansive agrarian economy. Patronage relationships between central rulers and religious institutions became the vehicle for creating infrastructure in the countryside, especially for land reclamation. Together, state and religious authorities built weirs and diversionary canals in well-watered areas like Kyaukse and Minbu. Authorities guaranteed cultivation and corvée labor in the area by legally tying laborers to the temple as part of the grant. Later this expansion took place on more marginal lands using various water capture techniques.[84] Through their water management and agricultural projects, religious authorities acted as proxies for the state, making this model more (but still not entirely) centralized. The key vulnerability of Bagan's system was the political relationship that made hydraulic expansion possible. Increasingly large private donations to Buddhist monasteries (made to avoid taxes) eventually bankrupted the state and contributed to the empire's collapse.[85]

Conclusion

The technologies that emerged in early historic Southeast Asia supported particular visions of flourishing life in increasingly complex societies. The advent of settled agriculture in Southeast Asia was accompanied by the emergence of more extensive material culture than that of hunter-gatherers and a characteristic pattern of technological dynamism spurred both by settlement and the formation of trade networks. Assemblages of technologies and techniques for construction, agriculture, water management, and specialist artisanal production supported new, more intensive forms of subsistence that served local populations and provided a secure foundation for trade. Trade, in turn, stimulated innovation in artisanal manufactures and facilitated the circulation of techniques and artifacts from China and South Asia. Intensified agriculture and trade played vital roles in the political development of Southeast Asia in the years before 1500, especially the emergence of the influential charter states. The technological dynamism of Southeast Asia is best seen by the human and material infrastructures coproduced with trade and urbanization. Those infrastructures would play a significant role in providing social, if not always political, resilience in the face of the challenges that would come into play in the fifteenth century and later.

CHAPTER 3

Textiles, Commerce, and Sociotechnical Resilience

DESPITE THE CONTINUATION of earlier trends such as steady increases in trade around the region, changing circumstances would reshape Southeast Asia's economic, environmental, and techno-logical character after the fourteenth century.[1] Areas formerly regarded as remote frontiers by larger polities grew in political and economic significance. The intensification and expansion of production and trade in valuable commodities affected both environments and economies. Sojourners and migrants, including European and Chinese traders, intervened in economies and politics to make their fortunes, introducing new techniques and modes of exploiting Southeast Asia's natural resources.[2]

These changes did not always produce economic benefits for Southeast Asians; some saw increasing wealth and others increasing poverty. Southeast Asian economies experienced periodic, regional declines in living standards after the fifteenth century and a long, slow decline in standard of living in the seventeenth century, part of the global seventeenth-century crisis characterized by political instability, population decline, famine, and warfare.[3] By the end of the nineteenth century, Southeast Asians experienced more poverty than during any previous period.[4]

Although the role of technological change in these declines has yet to be fully explored, the history of textile manufactures and trade offers a valuable lens through which to understand the relationship between technological dynamism and economic change in the early modern period. Textile manufactures and trade were central to cultural and economic life around the region. They also demonstrate the pressures that Southeast Asian manufactures came under as patterns of trade shifted in response to increasing interventions by foreign actors. Yet we should not view economic decline as a proxy for technological stagnation, incompetence, or inadequacy. As the case of textiles shows, technological change might contribute to impoverishment or social marginalization in some cases while serving as a source of resilience in others. Despite the economic dislocations of the period, traditional craft skills that underpinned textile production and technological innovation became a source of resilience.

Textile Manufacturing before 1400: Foundations of Resilience

Local production and long-distance trade in textiles were well established across the region at the cusp of the modern period in 1400 CE. Southeast Asians were increasingly importing both common and luxury fabrics from India and China, using the proceeds from commercial production of valuable crops like pepper and trade in Southeast Asia's forest and ocean products. Abundant cloth imports of silk and cotton fabric and thread are an important index of the general prosperity enjoyed in the period, thanks to the extant documentation of the price, quality of goods, and overall size of the trade.[5] Yet even extensive textile imports did not diminish the ubiquity of spinning, weaving, and other textile production skills in the region. Southeast Asians' ability to respond to disruptions from warfare or monopolistic power plays in the sixteenth century and later was based on the household and commercial skills established by the fourteenth century.[6]

The technological skills required to manufacture textiles are diverse. Depending on the fiber involved, it might include preparation like combing cotton or extracting fiber from the source plant, techniques for spinning or reeling thread, dying, and weaving. Sources of information on these technical aspects of Southeast Asian textile production are limited, even after 1400 CE, when written sources become more widespread. Much evidence is indirect and focused on trade, which can speak to the location of centers of technical expertise. At the same time, tax records and sumptuary laws specify different local categories of craftworkers and their products. Another indirect source of data is vocabulary. Words that reference spinning, weaving, and other practices offer insight into the kinds of activities that were prominent enough to leave a mark on the linguistic heritage of the region. Some technical practices survived to be documented later, although it may not always be easy to know how much they might have changed in the interim. Likewise, careful consideration of contemporary methods of spinning, weaving, and dyeing offer insight, if not absolute certainty, about the nature of earlier technologies. All of these indirect sources can help build a picture of the distribution and technical character of textile manufacture around the region at the start of the early modern period and offer insight into the ways that these technological foundations could serve as a source of resilience in the face of later economic change.

Cloth-making in Southeast Asia included woven cloth, typically made by settled communities, and nonwoven bark cloth, mainly made by nomadic or seminomadic communities. The making of bark cloth was an ancient practice in island Southeast Asia and involved a process similar to paper-making. The inner bark of certain trees, especially mulberry trees, was first softened by soaking in water, then beaten into sheets for several days and dyed or painted (figure 17).[7]

Such cloth was not durable—exposure to water would eventually destroy it. Therefore it tended to be used for rituals and other special occasions rather than for everyday wear. Bark cloth is still made for such events in some parts of Southeast Asia.

Before the ninth century, woven textiles were commonly made from the fibers of local plants, including hemp (*Cannabis sativa*), abaca (*Musa textilis*), lemba (*Curculigo latifolia*), pineapple (*Ananas comosus*), and sago palm (*Metroxylon sagu*). Although some of these fibers are still used

Figure 17. Ceremonial bark cloth costume from the Lore Bado people in the Regency of Poso in Sulawesi, Indonesia.
By 26Isabella via Wikimedia Commons. CC BY-SA 4.0, https://creative commons.org/licenses/by-sa/4.0/.

today, cotton and silk became the most widely used fibers for woven cloth in Southeast Asia by the 1400s.[8] At that time, Chinese merchants had long exported silk fabrics and reeled thread to Southeast Asia in exchange for forest products. The skills involved in silk production (mainly raising silkworms, reeling, and weaving) may have found their way to Southeast Asia along the same route, but there is little reliable information to confirm this. On Java, for example, silk garments were being manufactured as early as the sixth century CE; silk was clearly circulating in Javanese markets no later than the eleventh century, based on information from tax records and vocabulary.[9] Some sources suggest the presence of silk production in the region of contemporary northern Thailand, close to the border with China, at roughly the same time.[10] Indian practices may also have influenced the techniques of silk production in Southeast Asia due to the presence of luxury Indian textiles on Southeast Asian markets.[11] The patterning of fabric depicted on Angkorian statuary suggests that Javanese silks were being sold on that part of the mainland by the thirteenth century as part of a significant Southeast Asian trade in luxury fabric.[12]

Cotton was probably imported into Southeast Asia in the ninth or tenth century CE from South Asia. Not only does Indian cotton production predate that in Southeast Asia, but linguistic evidence from Java also suggests that much of the vocabulary referring to cotton and cotton production was derived from Sanskrit. By the eleventh century, the Song dynasty recorded imports of cotton textiles from Java.[13] Although silk remained a desirable luxury fabric, cotton cloth offered comfort and durability in the hot, humid climates of Southeast Asia. By the 1300s, Southeast Asians had taken up cotton cultivation themselves. Cotton was grown extensively in the state of Bagan (Myanmar); in East Java, Sulawesi, Bali, and Sumbawa (all islands in the modern state of Indonesia); and in Luzon and Cebu (islands in the Philippines).[14] In some Southeast Asian markets, cotton cloth was so desirable it became a form of currency.[15] On Java, by the 1400s, fabric gifts to high-ranking people might be made of cotton as often as silk (figure 18).[16]

Javanese sources offer particularly detailed information about the development of skills and techniques in textile production before the 1500s.[17] There, even ordinary women added to their daily repertoires domestic skills associated with making cotton cloth, including techniques for combing, spinning, and weaving. The sale of small textile tools like bows (for separating fibers), swifts (for winding finished yarn), spindles, dyes, and mordants in Javanese markets provides evidence of this change to domestic life.[18] Professional weavers, spinners, reelers, and dyers all existed by the twelfth century, probably specializing in finer fabrics.[19] Few studies offer similarly detailed insights into textile production in

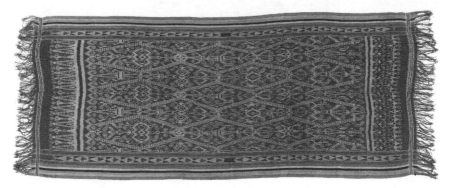

Figure 18. A Dayak ceremonial cloth made from cotton.
Courtesy National Museum of World Cultures via Wikimedia Commons. CC BY-SA 4.0, https://
creativecommons.org/licenses/by/4.0/.

other parts of Southeast Asia in the same period, especially those areas
not regionally recognized as centers for commercial production. Yet indirect evidence regarding trade suggests that domestic production of
everyday textiles remained common around the region.

The task most widely documented in texts, state records, and images
is weaving. The most common form of loom used across Southeast Asia
for everyday textiles was the backstrap, or body tension loom (figure 19).
The backstrap loom allows the weaver to anchor one end of the warp
threads to a strap across the weaver's back and the other end to some
fixed object.[20] Backstrap looms can be used with any type of fiber, including cotton and silk, although some modifications, including the use
of a comb or reed, were necessary to keep fine silk thread from getting
tangled.[21]

Backstrap looms are highly portable, making them useful for busy
households. Yet fabric woven on backstrap looms can be no wider than
the length of the weaver's arm; sarongs or other cloth that needed to be
wider than this had to be made by stitching lengths of material together.
By contrast, the fixed looms used in South Asia's textile industry could
produce wider fabrics. The appeal of such products for making seam-
less garments made Indian textiles especially desirable in the Burmese
market (and probably elsewhere).[22] In Java's commercial textile centers,
professional weavers called *cadar* worked on fixed looms, producing

Figure 19. A Kayan woman using a backstrap loom in Myanmar.
Photo by CEphoto, Uwe Aranas via Wikimedia Commons. CC BY-SA 4.0, https://creativecommons
.org/licenses/by-sa/4.0/.

finer-textured fabrics in cotton and silk than household weavers could typically make; the word "cadar" also refers to gauze.[23]

Cham and Javanese weavers made the most highly regarded cotton fabric in the region, although their best products never eclipsed the value of Indian cottons.[24] However, Southeast Asian techniques for sophisticated patterning and dyeing made these local cottons competitive.[25] Javanese inscriptions associated with sumptuary laws indicate the presence of new decorative techniques by the late twelfth century. Although it is difficult to be sure, these emerging techniques may have been precursors to Java's wax-resist batik technique, in which artisans draw sophisticated patterns in hot wax and then dye the cloth in multiple stages to achieve the desired decoration.[26] Southeast Asian dyers innovated using both local plants, which created a palette of earth tones, and newly cultivated imported plants like safflower for brighter yellows and indigo for deeper blues.[27] Until the fifteenth century, Southeast Asian dyes tended to fade quickly. Stimulated by imports of colorfast textiles, Southeast Asians eventually produced their own more durable dying techniques.[28] By around 1400, Southeast Asians had developed manufactures capable of producing a wide range of types of cloth, suitable for ordinary and luxury markets alike.

The Culture and Economy of Textiles after 1400

Although evidence shows that both men and women in Southeast Asia were engaged in various aspects of textile production, women were the dominant presence in most textile-related crafts, whether silk or cotton textiles, homemade or commercially produced. For example, in Surakarta in central Java during the seventeenth and eighteenth centuries, thousands of women typically engaged in textile production. Some of these women worked out of their households (usually for the production of ordinary cloth and often during the agricultural off-seasons). Others worked in *dalems*, workshops operated either by Javanese nobility or Chinese merchants.[29] In the Malay world, many women regularly participated in commercial activities and enjoyed high status in society.[30] Many weavers marketed their own cloth (as other women marketed agricultural commodities) and exercised considerable economic and social agency.[31] It is less easy to document their role (or any specific person or group's role) in technological innovation. Still, it is likely, given the demographics, that women were responsible for many notable innovations in cloth-making in the early modern period.[32] In other parts of Southeast Asia, women's roles were different, but household cloth production was also carried out mainly by women.[33]

This gendering of production was no accident. In Malay cultures, cloth was assigned female qualities of life-giving and protection, ideas that were present in other parts of Southeast Asia as well.[34] Finely made textiles were prized gifts at the highest levels of society. Aristocrats used luxury textiles to affirm important political alliances not just because of the cloth's significant economic value but also because of its cultural meanings. Commoners also marked important occasions such as marriages and births with gifts of cloth. In the Malay world, aristocratic ladies wove cloth with their own hands for special gifts or rituals, imbuing these textiles with special significance.[35] Sumptuary laws restricted certain colors and fabrics to those privileged to wear them, with yellow frequently being restricted to royalty. Despite many differences in particulars, Southeast Asian cultures shared a fundamental view of the cultural significance of textiles.[36]

The growing prosperity of many Southeast Asians in the early modern period offered real incentives to import India and China's fine cottons and silks. In Angkor, although some patterned cloth depicted on statuary appears to have been from Java (as mentioned earlier), others have the trademarks of imported Indian silk.[37] Although they cultivated cotton in commercial quantities in Burma, elites and commoners alike wore textiles imported from across the Bay of Bengal. Cloth's meaning and prestige contributed to the desirability of imports. For example, in the seventeenth century, increasingly wealthy pepper farmers from the inlands of Sumatra developed a taste for imported cloth, including fine cottons and silks, that had previously been available only to aristocrats. When political elites no longer monopolized the finest imports for their own use, as happened in eighteenth-century Banjarmasin, they became anxious that the loss of sumptuary distinctions would destabilize royal authority (figure 20).[38]

Because pepper-growing was a particularly time-consuming task, royal authorities worried that fewer women were weaving—a justifiable concern under the circumstances, and one that suggests that weaving itself, and not just the cloth it produced, was connected to ideas about social order.[39] Yet despite the availability of desirable imports, locally produced textiles never wholly vanished. Weaving remained a site of woman-centered technological skill and business acumen through much of the early modern period. The retention of necessary skills despite the presence of imports can be seen by examining what happened when European-based traders entered the cloth market in the early modern period, extensively disrupting established mercantile relationships and the flow of affordable cloth around parts of the region.

Figure 20. Woman weaving songket, a luxury fabric that usually included gold or silver threads, in Surabaya, c. 1905.
Courtesy Leiden University Digital Collections. Shelfmark KITLV 10836. CC BY-SA 4.0, https://creativecommons.org/licenses/by/4.0/.

Technical Skills and Resilience

Europeans, in particular the Dutch East India Company (VOC), sought to establish trade monopolies on spices and found that the product they could most easily use to obtain what they wanted was fine Indian cotton and silk cloth.[40] As the Dutch worked to monopolize Indian cloth trade, prices rose, a problem intensified when warfare on the Coromandel Coast (in southeast India) in the seventeenth century limited cloth production, creating both price and supply shortages.[41] On Java, for example, cloth imports, considered by many economic historians an important indicator of overall prosperity, declined by 90% during this period.

The presence and resilience of artisanal communities played a significant role in allowing local polities to quickly compensate for the loss of imports. The maintenance of technological skills in weaving, tied to women's domestic and commercial lives, was a key reason that the decline in cloth imports was far from catastrophic. In the Malay world, the continual demand for cloth of diverse kinds and qualities meant that local weavers had continued to make cloth for everyday use, even if many people aspired to obtain higher-quality imported fabrics, too. Artisans had also continued to improve the quality of local cloth over time, including the introduction of colorfast dyes and preshrunk cotton threads.[42] When Javanese weavers turned to the production of luxury fabric in silk and cotton, representatives of the VOC admitted that it was nearly as good as Indian cloth in many respects.[43] The very speed of the switch from Indian to local cloth allows us to reasonably speculate that weaving skills had not declined even when imports of Indian cloth were high.

The cultural meaning of cloth and its significance for prestige and display also stimulated local manufacture during this period of disruption. Evidence of this comes from the royal courts of Palembang and Jambi (kingdoms on the island of Sumatra, which is located in present-day Indonesia). Although they represented a vanishingly small proportion of the women involved in textile production, as both artisans and marketers, noblewomen were agents for change in the textile trade.[44] When Indian cloth was unavailable or offered at exorbitant prices, they would sometimes run smuggling ships to obtain the cloth they needed. When imported fabrics failed to meet their demands for quality and design, the women of the courts wove cloth to their own specifications.[45] Nonroyal women attached to these courts spread the skills and techniques required to make such fabrics further into ordinary society, albeit using less expensive fibers and with an eye to the restrictions of sumptuary laws. The prestige of the Palembang court nevertheless gave this local, nonroyally produced fabric a cachet that must certainly have attracted imitators.[46]

Although local cloth manufacture provided a source of economic resilience in the face of VOC manipulations, for ordinary women weavers, the rapid changes came at a price. The growing role of foreigners, especially Chinese merchants, in funding the expansion of textile production may have diminished women's traditional social and economic power and devalued the very technological skills essential to responding successfully to the VOC. Chinese (and some Indian) merchants offered credit to help women finance expanded production.[47] Fast-growing weaving centers that emerged in places like Semarang, Gresik, and Surabaya (all on Java) were so successful they even began to export cloth.[48] However, to achieve these gains, single merchants might put all the weavers in a village under contract, a marked contrast to the traditional system where individual women marketed their own products separately. Women's economic independence accordingly decreased as they lost the ability to negotiate acceptable prices on their own behalf, as well as the power to decide what products they would make.[49] Foreign merchants, unaccustomed to dealing with women as primary decision-makers, treated women with less respect than they traditionally received in Malay society, profoundly disrupting the gendered social order in the region.[50]

In the realm of dyestuffs, the growing demand for indigo likewise affected women's traditional roles as growers of indigo (a crop used to supplement household income). As indigo moved to estates operated by Chinese or European owners, male laborers took over much of the work, except that which they considered demeaning, like weeding. Consequently, women's work in the hardest field labor increased while their presence in markets decreased.[51] Growing tensions between ethnic Chinese and indigenous populations in the Malay world, which would plague the region for centuries, are evident from this time and were undoubtedly exacerbated by conflicts about women's roles and power in society.[52]

Conclusion

This exploration of textile industries highlights the ways that technology and technological change were embedded in contests for power during the early modern period. The desire to imitate or fill a market niche left open by the absence of Indian cloth spurred local innovation in both cloth-making and labor organization. Javanese weavers and entrepreneurs, in particular, responded so quickly to fill the gap that little hardship due to the loss of imported fabric was evident. Local royalty and merchants alike swiftly moved to make the quality cotton and silk textiles that were increasingly difficult (or undesirable) to buy.

Although Southeast Asians could develop and use technological skill as a form of economic resilience, the pressures applied by the VOC and Chinese traders on the sociotechnical order of society was significant. The VOC hoped to benefit strictly by manipulating the market through monopoly, while Chinese traders opportunistically reorganized labor and production to ensure their control over desirable trade goods. Although artisans such as weavers and dyers are only indirectly visible in many primary sources, it is clear that artisans and their skills were central to attempts to alter interdependencies and thus power relations.[53] Technological skill, far from disappearing in the economic disruption of the period, became a fulcrum for shifting power in the region.

CHAPTER 4

Localizing Foreign Technology

Mining and Shipbuilding in Early Modern Southeast Asia

AS IMPORTANT AS long-established technologies and techniques were in the early modern period, knowledge and technologies introduced from China and Europe played an equally prominent role in the period's economic and commercial dynamics. This chapter explores Southeast Asian strategies to localize foreign mining and shipbuilding technologies, deploying them to benefit Southeast Asian actors. Mining and shipbuilding offer perspectives from opposite poles of the spectrum of localization practices. Mining technology in Southeast Asia was transformed into something nearly identical to the models developed in Chinese mines. Shipbuilding, by contrast, was characterized by mutual learning among Southeast Asian, Chinese, and European actors, producing technologies whose character cannot be easily attributed to just one of these groups.

Both cases highlight how circulating technologies were implicated in emerging forms of sociotechnical interdependence between Southeast Asians and people from outside the region. Forms of technological activity such as mining and shipbuilding grew dramatically in commercial value as local and foreign skills, labor, and knowledge intermixed. They stimulated and reinforced patterns of migration and sojourning that, in turn, transformed cultural and political relationships around the region.

Mining: Chinese Migration in the Logic of Technological Intensification

Although mining operations that extracted gold, silver, iron, copper, and gems stretch back to antiquity in sites across Southeast Asia, the period after 1400 saw a dramatic change to mining's political economy and technological practices. Especially in the long eighteenth century, mining operations in mainland and island Southeast Asia intensified dramatically, driven by growing demands for metals and gems in Chinese markets. Intensification is evident in increases in the ore production at a given site and in the amount and kind of labor expended to produce those results. Mining intensification was made possible by the extensive migration of Chinese laborers and mining entrepreneurs across the region.

Labor-driven intensification could only be achieved via migration (voluntary or forced) because Southeast Asia's populations were sparse. Therefore, localizing practices that brought labor-intensive forms of Chinese mining to Southeast Asia resulted in broader social transformation via the creation of migrant communities. The integration of Chinese migrants and communities into their host polities varied significantly from place to place. In Siam and the Philippines, Chinese migrant communities ultimately assimilated into local cultures through intermarriage. In the Malay world, politically enforced ethnic separation resulted in long-term social tensions and unique new cultures like the Peranakan Chinese. In Yunnan, long-term, large-scale Chinese migration fell just short of full sinicization, the establishment of a dominant Chinese culture at the expense of the region's indigenous people.[1] Therefore, the presence of mines and the nature of their operations carried significant consequences not just for the economics and environments of Southeast Asia but for their social histories as well.

Mining before the Eighteenth Century

By the sixteenth and seventeenth centuries, Southeast Asian states imported considerable amounts of copper and iron from China in the form of finished goods and ingots. The reason for importing rather than mining and smelting their own came down to cost and the limited availability of labor. Southeast Asian methods for both mining and smelting were more expensive than Chinese methods and required considerable labor, which was not always available.[2]

In several important areas of central Southeast Asia, particularly the islands of Karimata and Belitung near Borneo, a long-established trade in iron goods, especially high-quality axes and knives, stretched back to at least the thirteenth century. There is evidence that Chinese artisans had settled in the region, creating a flourishing trade with Javanese and Balinese kingdoms and the newly arrived VOC. However, it is not clear what techniques were employed, whether Chinese, Southeast Asian, or some hybrid of both. Over time, Karimata producers were pushed out of the market by cheaper imports from China.[3]

Before the introduction of Chinese mining in the region, the political economies of mining varied from place to place. Several Shan principalities in what is now Myanmar took advantage of a lucrative Chinese market for gems during the Ming dynasty (1368–1644) to expand their power and influence, although there appears to be little information about their mining practices.[4] Yet mining was clearly central to their growing political strength.[5] By contrast, the stateless Wa people (who lived in both Yunnan and what is now the border region between Myanmar and China) used mining to help them resist political centralization. They operated lucrative silver mines using enslaved laborers and later imported Chinese miners and mining practices, maintaining Wa fiscal interest in the mines.[6]

In other areas, mining was pursued as an ancillary economic activity as labor availability permitted. In Sumatra, for example, people in farming communities who provided rice and other food to nearby cities would pan for gold in rivers during the agricultural off-season.[7] On the island of Borneo, regional leaders looked to seminomadic foragers, the Dayaks, to provide gold in exchange for goods like rice, salt, and opium.[8] Their techniques were likely the same panning as that practiced elsewhere in the Malay world; the Dayaks themselves controlled the pace of extraction.

Because metals were essential for producing weapons and coinage, authorities frequently opted to exert more control over mining operations. They required some portion of the proceeds of mines to be returned as tax or arranged guaranteed purchases with private mining operations, as was the case in Yunnan under Chinese leadership.[9] However, particularly on the mainland, much mining took place in areas distant from centers of power. As mining became essential to states, authorities worried (with some justification) about the political reliability of mining regions. Many mining areas were located on sometimes poorly defined borders. Mining areas often became sites for political unrest and outright rebellion, as in northern Vietnam and Yunnan.[10]

The Circulation of Chinese Mining Techniques to Southeast Asia

Change came to Southeast Asian mining largely because of the challenges faced by China's Qing dynasty (1644–1912) in securing an adequate metal supply for coinage. The Chinese population and economy had seen tremendous growth since the Ming dynasty (1368–1644). The extensive use of coins in the Chinese economy required metals, especially copper and silver, to satisfy the need for currency. Imperial authorities had determined to revert from paper to metal currency for good reason: minting coins was a profitable activity.[11] During the Qing dynasty, with Japan's Tokugawa Shogunate restricting the metals trade China had earlier relied on, imperial authorities sought new sources of metal and expanded existing production.

Some of the richest lodes of copper and silver were along the present-day border between China, Vietnam, and Myanmar, as well as in Yunnan. Although the Chinese market and tributary relationships stimulated mining across this region, Chinese authorities made territorial claims on Yunnan. It was here that the most dramatic expansion of copper and silver mining occurred. Located in a region that is now part of China, it was a true borderland in the eighteenth century. Although the Chinese had claimed authority over parts of the region since at least the Han dynasty, in practice, they had only sporadically enforced their claims.[12] Tai peoples with a strong cultural orientation aligning them with Southeast Asia were indigenous to the area.

Chinese migration and colonization expanded dramatically as soldiers, farmers, and miners flocked to Yunnan during the Qing dynasty, particularly under the Yongzhen Emperor (r. 1722–1735).[13] The result was less a thoroughgoing sinicization (or conversion to Chinese culture) than a complex intermingling of Sinic and Southeast Asian cultures, politics, and societies.[14] Although the historical record seems to first take note of Yunnanese mines in the late 1730s, indirect evidence suggests that mining operations had been underway since at least the late 1600s.[15]

The methods of working mines in Yunnan and other border regions dominated by Chinese technological actors were generally based on labor-intensive Chinese practices. Given the demographic realities of China, already densely populated by this time, such methods were easily scalable in a Chinese context. The best description of Chinese mining methods of the time comes from Wu Qijun's *Illustrated Account of the Mines and Smelters of Yunnan.*[16]

A mine master who had the capital and knowledge to organize production would be in charge of the mine. This boss would hire laborers (known as "brothers"), who received food and tools but no wages. For large operations, major investors (merchants) owned shares of the mine but were otherwise uninvolved with daily operations. Everyone, from investors to laborers, received a share of the proceeds. The potential for good earnings attracted the poorest laborers to the dangerous, backbreaking work required of them.[17] This system was flexible, accommodating both large and small operations. Small operators could work mostly played-out old mines; others subcontracted to work small areas of more extensive operations. All adhered to the general organization described previously.

Chinese mine masters used empirical methods to prospect for metals, exploring characteristics of terrain associated with particular ores. These were useful methods, if inexact; many mines failed quickly.[18] Once the master located a potential mine site, workers dug shafts (either horizontally or vertically), inserted supports to keep them from collapsing, and then ventilated them with bellows. Removing water was even more technically challenging and financially demanding for all but the shallowest mines. In Yunnan, miners reputedly used lift pumps (suction pumps), possibly originating from Europe, to clear water from the mines.[19]

Chinese-style pumps, such as the square pallet chain pump normally used by farmers, worked in other locations, but they were a poor fit in Yunnan. The mountainous terrain made running the water wheel that typically powered such pumps difficult. Lift pumps worked better but required at least three, and as many as six, men each to operate. A typical mine required about 100 pumps. For large mines, a *xiangtou*, a sort of mining engineer, designed ventilation, mine supports, and drainage. Miners worked in teams of three, cycling through a work-rest cycle: one to wield the hammer, one to hold the chisel, and one to rest. Others hauled the heavy ore out of the mine.

Smelting happened on-site, and here, too, experts were essential for a good result. Furnace masters supervised the smelting in single or sometimes chains of multiple furnaces, which resulted in a purer product when used together. Chinese furnace masters were in high demand in Southeast Asia, even when Chinese miners were not employed, for example, in parts of Thailand and the Malay world, because of the good results compared to indigenous smelting.[20] That being said, archaeological investigations of mine sites show up to 5% of the ore remaining in the slag of some Yunnanese mines.[21] This inefficiency, combined with charcoal as a fuel source, had significant environmental repercussions. Severe deforestation transformed the Yunnanese environment around the mines.

This system of mining required considerable labor if practiced on a large scale, yet this hardly presented a problem in Yunnan. Because mine masters could offer landless people, or those with poor lands, the promise of an income, they could attract plenty of Chinese workers to this remote region, especially as imperial authorities strongly incentivized migration. Numbers are hard to calculate with any precision as not all miners would be registered with the central government (only landowning families could register as full citizens). However, miners in Yunnan probably numbered in at least the tens of thousands by the middle of the eighteenth century.[22]

This model of mining moved into other parts of Southeast Asia through the agency of sojourning Chinese merchants and technical experts. They brought to Southeast Asian polities not only experience and technical skills but the ability to mobilize immigrant labor for their intensive operations, a precondition for any successful operation on this scale in sparsely populated Southeast Asia.[23] Consider, for example, the mines of Bangka and Belitung in the Malay Archipelago, two islands that house some of the world's richest tin ores. As tin's export value increased in the late eighteenth century, the Sultan of Bangka increasingly employed Chinese miners to work the tin fields.[24] A *tikus*, often a half-Malay, half-Chinese intermediary, would be given permission to open a mining operation. They would team with a mine master who would recruit laborers (and keep recruiting, as needed). Thousands of Chinese miners made their way to the tin fields of Bangka and Belitung this way.

It is unclear whether mining expertise, especially the work of the xiangtou, passed into the hands of non-Chinese local peoples. Mines were usually located in remote rural spots, which in some cases could become enclaves of mainly Chinese settlers. For Southeast Asian peoples, mining labor likely held little appeal. For some, such as the Wa mentioned previously, mining was a culturally undesirable activity, suitable only for enslaved people or foreigners.[25] Other Southeast Asians generally had better options than the harsh labor of mines. Intermarriage between Malay and Chinese people certainly did occur. But some Chinese men who married local women left mining entirely and instead took up the business activities of their wives' families. Little evidence suggests it worked the other way, with indigenous families taking up skilled mining work, although more research might shed light on this matter.[26] Legal restrictions, too, could prevent the deeper assimilation of Chinese miners into local societies. Chinese knowledge, therefore, seems to have stayed primarily in "Chinese" hands, even if some of those hands were actually of joint Southeast Asian and Chinese heritage. The Malay saying "Where you find tin, there you will find Chinese" speaks to a technopolitical history of ethnic segregation that was a hallmark of the tin-mining regions, a co-constitution of technology and plural societies that has not yet received the scholarly attention it deserves.

To localize Chinese mining practices to Southeast Asia required trade and political relationships and a steady flow of Chinese labor, skilled and unskilled. Although the stable techniques and technologies of mining did not undergo much change in Southeast Asian settings, deploying this system of mining nevertheless required considerable sociotechnical change on the part of Southeast Asian societies. In particular, the system hinged on an uninterrupted flow of people into and around Southeast Asia, creating in some areas technological enclaves unlike anything previously experienced by Southeast Asians. The interdependency that it created—between Chinese and Southeast Asian actors each seeking benefits from the mining industry—powered a profound cultural transition in mining regions as Chinese workers brought Chinese ways of life and family connections that continue to shape those areas up to the present day. Although rich histories of Chinese migration do exist, there has yet been little attention to the sociotechnical dimensions of immigrants' lives and the ways that Southeast Asian technological cultures changed as a result. Further research would offer valuable insight into the relationships between technology and ethnicity in Southeast Asia.[27]

Shipbuilding: Skill and the Logic of Hybridization

The role of foreign technology and techniques in early modern shipbuilding offers a sharp contrast to the enclave-centered history of mining. Shipyards were intrinsically cosmopolitan places. As traders made their way to Southeast Asian ports to do business, so did their damaged ships. They eyed each other's designs, engaging in mutual learning and hybridization of different traditions of ship construction. The presence of new players, especially European trading companies, and the realities of conflict within Southeast Asia stimulated attention to shipbuilding and ship design as a resource for gaining or retaining advantages during turbulent times.

Shipbuilding highlights the developing technological interdependence between Southeast Asians and non–Southeast Asian sojourners who flooded the region in the early modern period. Southeast Asian shipbuilding was historically consequential within and far outside the region.

Spanish galleons, for example, relied on Philippine shipbuilding expertise to maintain their transpacific empire into the nineteenth century; China's Qing dynasty looked to Vietnamese shipbuilders to improve China's navies.[28] This perspective fits recent historical trends that explore both conflict and interconnection in histories of Southeast Asian trade. European and Chinese sojourners depended on the robust foundation of Southeast Asian infrastructure to make trade and colonial enterprises work. Southeast Asian shipbuilders localized foreign knowledge, building materials, and intellectual infrastructures that kept trade strong.

Hybrid Ships

Hybridization in shipbuilding included changes to design and function in the inventory of ships that sailed the region and the less visible work of bringing local techniques and materials to bear on imported ship designs to make them more functional in Southeast Asian seas. The Manila galleons, which carried Mexican silver to Asia, demonstrate the localization of European ship designs.[29] Spanish ship captains took advantage of local shipbuilding expertise to solidify their place in Southeast Asian trade (figure 21). The Pacific passage was unkind to Spanish ships, which often arrived heavily damaged and in need of a complete refit. Like other European ships, Spanish ships were built of oak and avidly attacked by shipworms in warm, tropical waters.[30]

The coastal peoples of the Philippines had significant shipbuilding expertise. They built many styles of ships, including the large *karakoa*, a warship outfitted with outriggers (figure 22). The Spanish were duly impressed with the design and sailing capacity of these ships. For refitting, they often hired local Chinese overseers and employed Philippine artisans as carpenters or in other work the Spanish viewed as less skilled. However, the skills and knowledge involved in successfully refitting a galleon were significant. They needed to find suitable local wood and other forest products (including resins and material for rope) and craft it appropriately. Even the search for wood could be a major undertaking. In one case, the Spanish used 6,000 local laborers to find and transport

Figure 21. A Spanish galleon in the Marianas Islands.
From the Boxer Codex (1590). Unknown Spanish author/compiler with likely Chinese artist from Spanish Colonial Manila, Philippines, circa 1590 AD. Public domain.

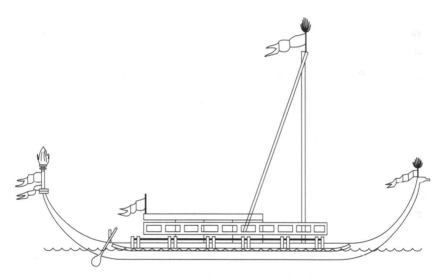

Figure 22. Drawing of the superstructure of a Visayan karakoa.
By William Henry Scott, "Boat-Building and Seamanship in Classic Philippine Society," via Wikimedia Commons. CC BY-SA 4.0, https://creativecommons.org/licenses/by-sa/4.0/.

the material for a single galleon.[31] Yet Spain relied on compulsory labor to keep their costs down, creating deplorable working conditions.[32] Rebellions occasionally forced the Spanish to have ships built elsewhere in Southeast Asia.[33] The low wages they paid and their extensive use of forced labor tend to obscure the reality that Spaniards depended on Southeast Asian skills and knowledge for their own fortunes. Therefore the globally consequential Spanish galleon trade rested on a hybrid infrastructure of European, Chinese, and Southeast Asian knowledge, skill, labor, and materials.

Elsewhere in Southeast Asia, the seventeenth and eighteenth centuries saw a shipbuilding boom spurred largely by Chinese commercial activities. In Siam and Vietnam especially, the China trade was a significant factor in the eighteenth-century expansion of shipbuilding prowess.[34] China's plans to import rice from Southeast Asia stimulated shipbuilding in this period. In the early eighteenth century, the Qing government encouraged the rice trade with Southeast Asia but refused to allow traders to deal in the more valuable goods they controlled. Rice trade was hardly an enticing prospect for Chinese merchants, given the cheapness of the product. To sweeten the deal, the Qing government allowed merchants to build ships in Southeast Asia and then transport the rice on the newly built ships, making the enterprise far more profitable. Chinese traders, therefore, sponsored an eighteenth-century boom in shipbuilding in Vietnam and Siam, kingdoms with ample rice to export.[35] As the quality of these ships became evident and the shipbuilding business took off, European traders also began buying ships for the same reason—to make a low-value cargo pay for itself through the quality and cost-effectiveness of the vessel in which it was carried.

By the early nineteenth century, some of the finest ships in the region were being built in Vietnam. European merchants may have asked for European-style designs or at least European-style features for the vessels they ordered. The expertise to build European ships or adopt aspects of European design probably came from interactions between Europeans and the diverse (Chinese, Cham, Viet, and other) populations of Cochinchinese shipyards of southern Vietnam. It is likely that the successors of these knowledgeable Cochinchinese shipbuilders later migrated to Chanthaburi in the nineteenth century, driving a considerable expansion of Siamese shipbuilding.[36] The growing demand for ships underscores the interdependence between successful foreign trade and the prosperity of Southeast Asian artisans.

Ultimately, Southeast Asian shipbuilders provided the backbone of technical skills that contributed to developments in naval architecture around the Southeast Asian mainland (see chapter 6). Long-term war-fare in Vietnam (and elsewhere) drove innovations in shipbuilding.[37] An important figure in the history of innovative shipbuilding was Nguyễn Ánh. He was a lord of the Nguyễn family, which had formerly ruled the southern portion of Vietnam before the ascendancy of the Tây Sơn dynasty based in the north. He led the opposition to the Tây Sơn in the late eighteenth century and later established the Nguyễn dynasty (in 1802) to rule the united nation of Vietnam, taking the imperial throne as Emperor Gia Long. As he fought the Tây Sơn in the late eighteenth century, he sought to shift battle to the sea, where his navy would have significant advantages. Using plentiful supplies of timber in the dockyards in Saigon, he sponsored a massive buildup of his navy, which ultimately included ships of both Asian and European design.[38] For example, the European schooner, a type of ship characterized by rigging the sails along its keel, rather than perpendicular to it, was included. The Tây Sơn built up their navy in response, thus incentivizing the growth and spread of shipbuilding skills.[39]

New designs were likewise motivated by interaction with foreign ships and shippers in the Malay world. Yet indigenous Southeast Asian crafts remained critical resources in the face of the expansion of both Dutch and British commerce in the region. In the seas around Java in the eighteenth century, the harbormaster records suggest at least 47 different ship designs in use, although it is not always possible to find a full description of such designs in other sources.[40] A common commercial ship type was the locally built *mayang*, a Javanese design characterized by a flat bottom, a single rudder, and a curved bow and stern with a single mast. Used both for fishing and carrying cargo, it was probably 30–40 feet long. The *gonting* used in Java and Makassar may have been a larger version of a mayang, without the curved bow and stern, although only poor descriptions are available. The *pencalang*, a workhorse cargo ship, was originally built in the region around the Straits of Melaka but by the 1500s was built in Java as well (figure 23). It was a 40-to-50-foot single-masted ship with a rectangular sail, a deck, and a lateral rudder.

European-style ships were also well represented. The *chialoup* (also shallop), a type of sloop, was originally designed in Northern Europe

Figure 23. A sketch of a pencalang from 1841. Note that it is mislabeled as a mayang.
From François-Edmond Pâris, *Essai sur la Construction Navale des Peuples Extra-européen*. Public domain.

but adapted for use in Southeast Asia by local builders. It had one mast and one deck and could be designed either with a single rudder, as in the European style, or with a Malay-style dual rudder. Another European design was the brigantine (or *brigantijn*, as it was known in Dutch), with a single rudder, two decks, and two masts, between 70 and 100 feet long. These ships were built in the Malay region (many at the Sulawesi shipbuilding cities of Bima and Bulukumba) by Javanese and ethnic Chinese shipbuilders.[41] Scholars have noted how difficult it is to disentangle these many ship types, which indicates how common it was to make continual, ongoing modifications to existing designs.[42]

Different ship designs predominated as the value of products and routes changed over time. Although the VOC played an important, sometimes economically depressing, role as they increasingly taxed locally built ships and attempted to enforce monopolies on the most lucrative trade, they did not control all private trade, especially with China. In this private trade, the most significant shifts in ship design occurred. In the early eighteenth century, when the Makassar-Maluku route (from the central to eastern portion of the archipelago) was important, the *chialoup* predominated. After 1750, however, that route and the number of chialoups in operation declined. Instead, the *paduwakang* became more prominent. The paduwakang, originating in Sulawesi, had two decks and two lateral rudders in the Southeast Asian style.

It was 25 to 65 feet long, although the smaller sizes were probably more common by the eighteenth century. Despite being a physically unstable craft due to a high stern and low bow, it accounted for 60–90% of traffic in this important trade route.[43] Its modest cost to build, carrying capacity, and crewing practices made it a more economical choice than the chialoup, an increasingly important consideration as the most lucrative cargos moved into the VOC's hands.

As we saw in the case of weaving, efforts of the VOC to control trade did not diminish local shipbuilding skills nor saturate dockyards with technological products built or designed in Europe (figure 24). Instead, an assortment of cosmopolitan influences gave shippers options in challenging economic conditions. It incentivized the acquisition and hybridization of shipbuilding techniques and designs by builders, serving as a source of resilience to changing economic conditions.

Conclusion

The cases of labor-intensive Chinese mining practices and the boom in shipbuilding demonstrate the role that circulating technological skill and knowledge played in the commercial changes in early modern Southeast Asia. In both cases, Southeast Asians localized foreign technologies in ways that highlight emerging interdependencies between foreigners and Southeast Asians. To localize Chinese mining techniques in sparsely populated Southeast Asia required an ongoing dependence on immigrant Chinese miners. Foreign and Southeast Asian actors alike depended on hybrid technologies, techniques, and the cultures of mutual learning that blossomed in cosmopolitan shipyards to angle for advantage and solidify their place in Southeast Asian commerce. Whether resting on

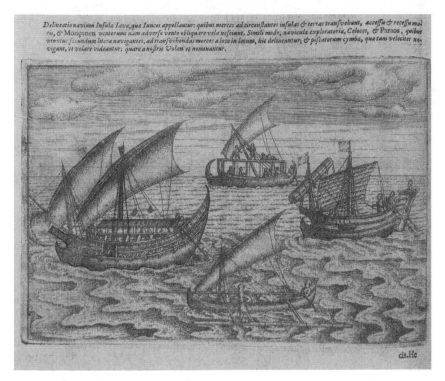

Figure 24. Sixteenth-century image of several Malay ship designs.
In Willem Lodewijcksz, *Prima Pars Descriptionis Itineris Navalis in Indiam Orientalem.* Public domain.

foreign or hybridized technologies, new dependencies could spark political tensions and violence, including violence against Chinese immigrant communities in the Malay world or rebellions against the demands of Spanish authorities in the Philippines. Yet violence was not the only consequence of these technological changes. Mining settlements played a role in the emergence of Southeast Asia's unique Chinese-Malay *peranakan* culture. The cosmopolitan shipyards arguably enhanced shipbuilders' technological agency and mobility as they found their skills increasingly in demand across the region.

CHAPTER 5

Intensification and Expansion

Agriculture in Flux

A S WITH THE OTHER TECHNOLOGIES we have examined, changing patterns of commerce and migration were the main drivers of change to agricultural practices in this period. For those areas most affected by Chinese and European sojourners and migrants, new crops were cultivated, and in some cases, farmers reorganized divisions of labor to accommodate growing demand.[1] Changing labor arrangements produced significant changes to social order and presaged the exploitative plantation practices that would become common throughout the region in the nineteenth century. Techniques to intensify production and the expanded use of preexisting techniques produced enduring environmental consequences that would shape the natural environment and agricultural possibilities in later centuries.

The Agricultural Transformation of Yunnan

Chinese migration, especially in the Yunnan region, resulted in new suites of agricultural practices and modifications to the Yunnanese environment to accommodate them. The Chinese imperial government motivated Chinese agricultural migration to Yunnan as part of a long-term campaign to assert control over the region and transform society and culture into something fitting dominant Chinese social, cultural, and economic norms. Numerous military campaigns attempted to pacify the regions; opening the land to agriculture was intended to provide an outlet and reward for military veterans and new opportunities for other Chinese farmers. Motivated by the appeal of tax-free land, nearly three million Chinese people migrated to Yunnan over the course of 150 years, although those numbers may not be completely reliable.[2]

By the eighteenth century, Chinese technological influences in this northern region of Southeast Asia were already pronounced. As early as the fifteenth century, Tày polities (located in what is now northern Vietnam) adopted Chinese tools and techniques in their rice production. The Tày peoples, a major ethnolinguistic group, lived in multiple distinct polities in what is now the border between China and Southeast Asia.[3] Tày peoples hybridized Chinese cultivation and Tày techniques and tools. They adopted the harrow and the plow from the Chinese, neither of which were previously in use. They combined these new tools, which largely displaced Tày hoes and harvesting knives, with Tày-style irrigation methods, including the use of bunded fields, weirs, and channels. Evidence strongly suggests transfer from Chinese to Tày farmers, rather than independent invention.[4]

In Yunnan, Chinese farmers likewise introduced techniques and tools unfamiliar to Yunnanese cultivators. More importantly, however, the Chinese conception of the best use of land was markedly different. Chinese approaches, therefore, represent more than just a change in techniques for familiar crops, as would have been the case in Yunnan. Instead, it introduced a system of intensive agriculture unfamiliar to Yunnanese and assumptions about wider agricultural economies that could not be easily harmonized with existing ways of life. Buckwheat was the most common upland crop in Yunnan before the eighteenth century, which was intended to complement hunting and foraging activities.[5] Chinese replaced buckwheat with sweet potatoes, peanuts, tea, and opium poppies in the fragile uplands. Chinese immigrants aggressively cleared land, using Chinese methods of fertilization to keep them productive. Chinese farmers viewed Tai pasturing practices as "wasteful," suggesting the strongly different perspectives they brought to the land. Communities that practiced hunting and foraging saw their animals and plants, important to their way of life, disappear as immigrant farmers permanently enclosed more land. Ultimately, many Yunnanese peoples took up Chinese practices, triggering a major environmental and economic shift around the region.[6] Although some Yunnanese managed to hold their traditional pastures, control their own villages, and thus maintain some degree of cultural sovereignty, the economy increasingly favored Chinese techniques for the intensive cultivation of cash and food crops. Ecologies, economies, and technological cultures transformed as a result.

Agricultural Expansion as Political Resistance? The Rice Terraces of the Ifugao

Foreign interventions also had indirect consequences for the suites of technical practices deployed for agriculture. The Ifugao rice terraces in the Philippines are a case in point. The Ifugao people live in the upland regions of central-northern Luzon. They are most famous as the builders and maintainers of the Banaue rice terraces, now designated as a UNESCO World Heritage Site (figure 25).

The terraces offer a striking example of labor-intensive landscape transformation, with flat bunded fields carved into the mountainsides to accommodate flooded rice cultivation.[7] Still in regular use, they host mixed cultivation of wet-rice and dry-season crops like taro. Before these areas were used for rice cultivation, they were dedicated solely to taro and other cultivars suited to dry conditions. Their transformation to rice cultivation dates to around 1650, when Spanish colonizers asserted their authority in the nearby lowlands.[8] What stimulated this change and justified the extraordinary labor involved in creating an environment appropriate to wet rice in this challenging environment?

One popular idea is that upland areas like the Philippine Cordillera became refuges for those seeking to escape the attention of oppressive lowland authorities, especially at this time the disruptive presence of the Spanish.[9] As the Spanish spread their influence through the Philippines, they pressed farmers to increase rice production, which the Spanish as-

Figure 25. Planting season on the Banaue rice terraces in the Philippines.
By Deanmanila via Wikimedia Commons. CC BY-SA 4.0, https://creativecommons.org/licenses/by
-sa/4.0/.

pired to sell. Rice was a desirable grain in precolonial times but not a
widespread staple.[10] The Spanish brought plows (and a monopoly on
their construction and sale) and introduced irrigation works around the
lowlands to raise rice yields.

That farm populations moved to the uplands to distance themselves
from colonizing Spaniards seems clear enough. But the Spanish still
taxed wet-rice agriculture; moving to the uplands provided no respite.
Taking up untaxed, commonly grown, and far less labor-intensive taro
would have made more sense. What better explains the extraordinary
effort involved in constructing rice terraces was the cultural significance
of rice beyond its mere economic value. Traditional rice production so-
lidified social and political bonds among the Ifugao people.[11] The rituals
and methods of working rice and harvest celebrations at which rice was
shared created strong social ties. Although the migrants to the uplands
shared significant ethnic ties with the upland peoples, rice production

may have helped integrate incomers with existing populations. Activities like communal work to build and maintain terraces, shared feasting when the crop was harvested, and numerous other traditional rituals surrounding rice, many of which were under siege by Spanish authorities in the lowlands, reinforced community bonds. Reproducing the wider culture tied to rice cultivation in the uplands had social value well beyond food provision. It materialized communal solidarity and signified opposition to Spanish cultural aggression.[12]

Expanding Production for Foreign Markets: Reorganizing Technological Labor

As was true in Luzon, demand for commodities on regional and foreign markets inspired significant changes in agricultural production in the eighteenth century and later. As was evident in Luzon, changes to tools or techniques were sometimes a part of this process, but not always. For example, when Chinese authorities sought more rice in the 1720s and 1730s, the Kingdom of Ayudhya (in present-day Thailand) and Vietnamese states all became commercial exporters.[13] They did so primarily by opening new areas to agriculture and providing farm labor for those regions, sometimes moving communities, and sometimes using captive labor on royal lands.[14] Preexisting systems for producing certain commodities were robust enough to underpin expansion if enough labor could be found.

At times, however, changing labor organization to meet new demands significantly affected wider sociotechnical systems of agriculture. Reorganizing labor fundamentally changed how agriculture was socially integrated into wider societies, and the consequences of these shifts could trigger changes to technical practices in turn. The introduction of gang labor–oriented plantation systems by Chinese and European traders and bottom-up changes to family labor organization shifted effort and energy to valuable export crops like sugar, pepper, and spices, with consequences for food and local commercial production.[15]

A case in point is the VOC's effort to monopolize the market in nutmeg, mace, and cloves, all produced in the Maluku area of Indonesia.

Although part of the VOC's strategy was always to control the flow of goods to ports and around the region, they aimed to completely control nutmeg production. After the VOC slaughtered nutmeg planters in the 1600s to restrict where and for whom nutmeg could be grown, they established the so-called *perkenier* system (where *perken* means something like "parks"). Enslaved laborers, including some Bandanese who had survived the massacre, cultivated tracts of orchards. They possessed silvicultural skills and knowledge essential to the enterprise, cultivating nutmeg trees underneath a canopy of larger trees to protect them from the sun, watering, weeding, and fertilizing them for best results. Although the VOC system produced more nutmeg than the Bandanese had done independently, this does not appear to result from any particular change to cultivation methods. Instead, the logic of monocropping replaced the mixed economy (which included food production) that had prevailed earlier (figure 26).[16] Although the specific methods of growing nutmeg do not appear to have changed significantly, the wider agricultural system was markedly transformed.

Controlling labor was equally important for Chinese and European merchants who tried to establish sugar production in the Malay world, the Philippines, Vietnam, and Siam.[17] For example, in the seventeenth century, European landowners rented land around the colonial city of Batavia on the island of Java (subsequently Jakarta) to Chinese sugar millers. The latter attempted to grow sugar by bringing in gangs of Chinese and Malay laborers. Over time, these fields declined in productivity because the sugar planters did not fertilize adequately—the transition to monocropping in the case of sugar required both labor, which they had, and knowledge about sugar's ecological effects, which they did not.[18]

Changes to divisions of labor in response to growing demand also occurred within families, as became evident during the expansion of pepper production in the seventeenth century.[19] Pepper probably arrived in Southeast Asia from India in the early first millennium. By the end of the seventeenth century, it was being grown in Sumatra, Java, Borneo, and Central Vietnam.[20] In the Malay world, families grew pepper to supplement household income, with women, who made most of the family economic and agricultural decisions, trading it at local markets.[21]

Myristicaceae.

Myristica fragrans Houtt.

Figure 26. Nutmeg.
In Hermann Adolph Köhler, *Köhler's Medizinalpflanzen*, 1887.
Courtesy Wikimedia Commons. Public domain.

Pepper is a crop that requires intensive attention and, therefore, labor. Adding pepper to a family's agricultural activities represents a significant sociotechnical change. After clearing land and planting pepper, vines require several years of careful tending before they will produce. Pepper vines deplete soil fertility drastically—if families intend to keep some land under pepper cultivation, they have to fertilize it intensively and continuously. Introducing pepper into a mixed agricultural system required experimentation with fertilizers and ongoing work that drew labor away from other crops. Grown in relatively small quantities, families could usually integrate pepper cultivation with existing commitments like rice and cotton, with only modest changes in their ways of working. Cultivated on a modest scale, it could even blend with the swiddening systems practiced by some upland people.[22] Given the value of pepper, it offered real benefits to cultivators so long as they could manage the labor requirements.

As demand for pepper increased and regional leaders pressed farmers to provide more pepper, the whole system of agriculture that supported these farmers became increasingly difficult.[23] In the sixteenth century, when demand for pepper grew, upland families responded relatively quickly, possibly because women substituted pepper cultivation for some of the labor traditionally spent on spinning and weaving. Pepper was lucrative enough that they could purchase good-quality cloth with the proceeds. However, demand continued to skyrocket, and regional authorities who had lucrative contracts with European and Chinese buyers pressured cultivators to increase production even more. For example, pepper cultivation in some parts of Sumatra increased 5,000% during the seventeenth and eighteenth centuries. With a sparse labor pool, compensating for the needs of pepper required bigger changes to agricultural practice.[24]

Because family size tended to be small, families with the means acquired either paid or enslaved workers. Yet the additional cost could reduce or even eliminate pepper's profitability for the family. Moreover, foreign merchants rarely traveled upland to the small markets where women usually traded, so women had to travel great distances to enjoy the highest income. Otherwise, go-betweens took a substantial portion of the profits. Women could not easily travel so far unless they had no children—the trips were far too dangerous.[25] With pepper commanding more of their time, women's control over family economics declined. This pattern worsened as European traders had little respect for women and frequently preferred to work with men.[26] Food production in some areas suffered, as some families chose to specialize completely in pepper.[27] In some areas, local rulers, appalled by the social change induced by pepper, insisted that pepper gardens be eliminated.[28] Yet European and Chinese commercial pressure and the lure of profits proved hard to resist. The problematic family economics and environmental difficulties did not eliminate smallholder pepper growing, but it did result in a major shift in the value of certain forms of technological activity compared with others. It undermined women's economic power in ways that may have affected the social respect their skilled work usually commanded.

An entirely different model for growing pepper arose in Riau on the coast of Sumatra in 1740.[29] Planters who specialized in the production of gambier began growing pepper as an adjunct to their production. Gambier, originally produced in this region by Malays, is derived from the *Uncaria gambir* shrub. Gambier had a host of traditional uses. It was used for medicine and chewing with betel nut, a mild stimulant common to the region. They also started using it for tanning leather by the mid-eighteenth century (figure 27). In 1740, Daeng Celak, the Bugis monarch of Riau, ordered the establishment of gambier plantations to enhance trade and pay for vital rice imports. The Bugis and Malay owners of these plantations recruited Chinese immigrants and overseers to work these plantations. By the 1780s, these plantations had added pepper to the rotation. Gambier, like sugar, must be processed rapidly after harvesting the leaves. Workers boil the leaves and filter them into a thick liquid, then pour the results into molds to harden. The leftover liquid from this process proved an effective fertilizer for the pepper plants. The value of adding pepper to gambier production may have become even more evident in the late 1780s, when warfare and turmoil in Riau resulted in the abandonment of Chinese plantations by the Malay and Bugis leaders who fled the area. The basic form of Chinese pepper and gambier production proved adaptable both politically and ecologically to other parts of the Malay world, especially Singapore. The joint production of pepper and gambier offered both a technical solution to the joint labor and environmental problems generated by expanding pepper production and an important adaptation to shifting economic circumstances around the region into the nineteenth century.

Conclusion

Technical changes to agricultural practice, whether through the expansion of familiar techniques to new areas or the introduction of new crops, had significant consequences for the social lives of people around

Figure 27. Interior of a gambier processing facility on a gambier plantation on Sumatra, c. 1880–1890.

Part of Album Or. 27.377. Courtesy Leiden University Digital Collections. CC BY-SA 4.0, https://creativecommons.org/licenses/by-sa/4.0/.

Southeast Asia and for the ecological landscapes in which they worked. Growing demand from Chinese and European markets stimulated much of this change, directly or indirectly. Women's social status was devalued in the Malay world as pepper cultivation spread to family holdings. In Yunnan, Chinese forms of intensive agriculture began to crowd out the pastoral practices of the Tai peoples native to the region. Plantations like the pepper and gambier operations in Sumatra and Singapore funneled wealth into distant hands and opened to permanent cultivation previously open land. Such trends would continue into the nineteenth century as European colonial control blanketed much of the region.

CHAPTER 6

Technology and Cultures of Conflict in Early Modern Southeast Asia

L IKE COMMERCE, warfare was a site of technological continuity and notable change in the early modern era. Endemic warfare, large and small, was far from unusual in Southeast Asia as polities fought for resources, honor, and security. Yet the growing presence of aggressive European trading companies, which were by their charters permitted to wage war on their own behalf, and occasional forays by Chinese empires complicated existing patterns of warfare.[1] Conflict, as much as cooperation, shaped circulations in Asia, and Southeast Asia is no exception.[2] As powerful actors vied for prestige, access to trade, and control over people and land, conflict produced changing ways of life, forced immigration, and technological circulation and change.

This chapter and the next explore the ways that warfare affected the technological lives of Southeast Asians in the early modern period, including continuity and change in the intensity of technological activity related to warfare and the consequences for the overall sociotechnical organization of their societies. Rather than focus mainly on military institutions, I explore how Southeast Asian societies integrated warfare into their sociotechnical lives.[3] How much technological effort was tied up in warfare or directly affected by violent conflict?

In warfare, as in so many other aspects of Southeast Asian history, finding meaningful, region-wide patterns can be difficult to do. Historians investigating early modern Southeast Asian warfare have sought to balance attention to its diverse local meanings and practices with region-wide trends, such as the adoption of firearms.[4] Balancing these two is essential for pushing back against misleading generalizations about warfare common in earlier scholarship. For example, it was long thought Southeast Asian warfare was fairly bloodless. Since population densities were low, the argument went, losing too many combatants would be self-defeating. That claim does not stand up to inspection in every case, although it may have been true in certain areas.[5] Recent histories that explore war and state formation, the consequences of the circulation of firearms for regional power relationships, and the interconnection between violence and trade have been careful not to overgeneralize while still seeking out regionally meaningful patterns.[6]

This chapter addresses the history of technology in a similar spirit, exploring local cases of technology use and circulation while attending to how these local technological experiences articulate with developments across the region. Local histories illuminate important moments of technological encounter—between kings and arms traders, peasant fighters and their elite comrades on the battlefields, between enslaved people and their captors, women and men, and between foreign actors and Southeast Asians. These encounters have much to tell us about the ways that technologies of warfare were embedded in social, political, and nonmilitary technological life.

The theme of scale is helpful for discerning regional patterns among local histories. I explore these military practices along a continuum rather than artificially separating warfare from "everyday" social violence in early modern Southeast Asia.[7] Fighting might occur between villages with a handful of combatants or between imperial armies numbering in the hundreds of thousands. Yet the experience of ordinary soldiers (and those who helped prepare them for battle) in these contrasting settings may not have greatly differed.[8] Exploring how Southeast Asian societies organized warfare at different scales offers insight into the intensity of warfare-related technological activity woven through Southeast Asian lives and how the sociotechnical practices of organized violence did nor did not change as warfare intensified. Investigating the sociotechnical nature of warfare at different scales and the responses of Southeast Asians to the pressures or opportunities of warfare highlights the complexity of technological change in the period.

Analyzing Technology and Warfare in Southeast Asia

While containing much of real interest, analysis of technology in the historiography of Southeast Asian warfare has been hampered by a legacy of deterministic thinking and Eurocentric framings of the research agenda. In the 1960s, literature that explored the material foundations of European trading empires, including the design of ships and weaponry like guns, was an important historiographic innovation. It brought attention to actions in Asia, rather than focusing solely on political disputes among European powers, even as it remained strikingly Eurocentric and oblivious to the wider contexts of Asian history.[9] Because of its inattention to the agency of Southeast Asians and the deeper context of Southeast Asian social and political worlds, technologically deterministic viewpoints prevailed. European technology was assumed to offer an unassailable advantage in regional conflicts.[10] Such analyses gave too much weight to technology without adequately contextualizing it, a problem exacerbated by a tendency to read European sources uncritically. At the same time, though, a sea change in Southeast Asian history was taking place, as more historians called for histories of Southeast Asia on its own terms. Careful explorations of the social and political histories of the region emerged. This trend has continued to the present, adding depth and nuance to all aspects of Southeast Asian history, including histories of conflict with Europeans in the early modern period.

Recent scholarship on Southeast Asian warfare, insofar as it pays attention to technology, generally rejects an overly deterministic view of its role in conflict. The oversimplification of complex social, political, and technological engagements inherent in deterministic analytics rarely stands up to the close scrutiny preferred by modern historians.[11] Although deterministic explanations can still be tempting when technologies seem to exercise an outsize influence on the outcome of conflicts, the move away from technologically deterministic explanations is certainly in keeping with the interpretive practices of the history of technology.[12] Analytically, however, rejecting determinism is just the first step toward a deeper understanding of technology and warfare in Southeast Asia.

Only one other theme emerges from the scholarly literature on technology in Southeast Asian warfare: the firearms revolution. Like technological determinism, this theme has Eurocentric foundations. Historians have argued that the embrace of gunpowder and firearms in Europe significantly changed not just warfare but also wider society.[13] Literature on the gunpowder revolution has spurred lively and productive debate.[14] Southeast Asianists have asked whether something similar happened there. Did the introduction of guns transform warfare and wider Southeast Asian society in distinctive and observably shared ways? Given the region-wide importance of firearms in the early modern period, it is a fair question. Most scholars who have explored this question closely find little evidence of a thoroughgoing or general military revolution in Southeast Asia, however.[15] In the context of technology history, comparison of the Southeast Asian experience with firearms to European military revolutions, while a reasonable enough undertaking, has not proven to be productive.

This chapter and the next attempt to offer a different entry point by thinking systemically about technologies of violence and probing how the design, use, and maintenance of these technologies were co-constructed with forms of social, political, and economic order. I do not claim that this is the only or best approach to pushing the analysis further. Instead, I offer it as a historiographic experiment, pulling material already available in existing literature to frame questions that focus less on effectiveness or comparisons with Europe and more on the changing ways that technologically mediated violence was integrated into Southeast Asian life. Such an approach can advance our understanding of sociotechnical order in Southeast Asia and provide a sounder foundation for historically contextualizing the role of technology in the advance of European territorial empires in the nineteenth century.[16] Doing so does not require that we ignore European or Chinese agency in the region nor discount the affordances of European- and Chinese-designed technologies. Still less does it require a return to the "essentialized" treatments of Southeast Asian warfare that scholars have so aptly critiqued.[17] Moving our attention to the character and durability of the sociotechnical effort devoted to conflict allows us to bring together in one frame production, use, and trade of traditional and new technologies, their ritual, practical, and social meanings, and engagements between foreign and local actors.

Violence in Southeast Asia in the Early Modern Period

In the years after 1400, endemic warfare and conflicts with extraregional actors, especially Europeans and Chinese, shaped the characteristic patterns of violence that emerged in the period. Conflicts, large and small, were nothing new. Intermittent small-scale conflict between villages or supra-village organizations is clear in historical and archaeological records, as are conflicts between larger polities.[18] Such warfare could be tied to the cycles of political fragmentation and consolidation that characterized the region's history from 800 CE into the nineteenth century.[19] Major conflicts that exemplified this trend include the rise and fall of the Taungoo Empire in Burma (1510–1752) and the Mataram Sultanate on Java (1587–1755), the rise of the kingdom of Makassar in the Malay Archipelago (1512–1654), the fall of the Kingdom of Ayudhya (1767) and the emergence of the Kingdom of Siam (1782), and the fighting by the later Lê (1533–1789) dynasts in what is now Vietnam and Cambodia. Therefore, the lives of many Southeast Asians would have been touched by organized violence in one way or another.

Many different concerns motivated warfare, including conflicts over resources, the desire to acquire slaves, the cultural need for men to prove themselves in battle, ambitions to expand political control over territory or trade routes, to settle matters of honor, or set hierarchical relationships of authority and prestige on new footings.[20] Such diverse justifications for warfare could translate into diverse practices, making it nearly impossible to define a characteristic "Southeast Asian" form of warfare. Small-scale raiding in some areas primarily involved the taking of heads, while in others destroying settlements and agricultural fields was a higher priority. Captives figured prominently in some cases, as when Burmese and Siamese armies sacked each other's capital cities to drive the demographic collapse of their opponents while adding to the populations under their own control.[21] Elsewhere captives were less central. Large-scale warfare might look like nothing more than extensive raids, as was the case when Đại Việt and the Cham Kingdom fought repeated battles before 1400, or a more tightly organized attempt at imperial conquest, as was the case when Đại Việt decimated Champa in 1471.[22] Warfare could be integrated into social and political order in markedly different ways, even when similarities in motivations or weaponry might be found.

In the early modern period, European and Chinese actors became increasingly relevant to the dynamics of technological change in Southeast Asian warfare. They supplied weapons and expertise, and their engagement with trade affected both the sites and character of conflicts. Chinese actors exerted the most influence in the borderland regions of mainland Southeast Asia, especially in Vietnam, where the Ming invasion of 1407 produced significant transfers of weapons, knowledge, and political organization that reverberated around the region.[23] European influence was felt most keenly along the coasts, as when the Portuguese attacked and ultimately defeated the rulers of the wealthy trade city of Malacca in 1511, and when the Spanish commenced their colonization of the Philippines in 1565 by establishing a foothold in the coastal city of Manila.[24] The VOC was perhaps the most widely active of all the European trading companies in Southeast Asia, using negotiation and violence to corner the market in the spice trade.[25]

Europeans were not responsible for all violence in the region. Still, their readiness to use force to gain access to trade added to the overall prevalence of warfare, especially as their interventions disrupted long-standing relationships. Southeast Asian leaders, for their part, were only too willing to recruit Europeans into long-standing conflicts.[26] Spanish, British, Portuguese, French, and Dutch agents also actively participated in the slave and arms trade. For Southeast Asians, European allies and captives became valuable sources of knowledge about European gunnery. At the same time, Europeans took up Southeast Asian tactics and technologies in conflicts. Over time, the imbrication of European justifications and means of war with that of the Southeast Asians meant that few disputes in the region can be easily understood, even technologically, in terms of "East vs. West." Alliances and advantages shifted as easily as guns and enslaved people were traded.

Small-Scale Endemic Warfare and the Technological Dynamics of Everyday Life

How much and what kind of effort was put into supporting warfare in all its dimensions in Southeast Asia? Consider how matters related to violent conflict intervened in everyday life for different peoples around the region. Small-scale warfare in Southeast Asia took the form of raids, where combatants aimed to gather captives, heads, or resources or to destroy their enemies' infrastructure. Such raids might involve territorial acquisition, although not always. For modern readers, taking heads may be an opaque motivation, but headhunting practice offers insight into how cultural meanings of warfare were co-constructed with technological practices. Headhunting was culturally important in many parts of Southeast Asia, including, for example, in the Maluku Islands, Sulawesi, and the Philippines, as well as in many regions across the mainland.[27] Frequently, the concept of charisma informed the logic of headhunting.[28] Leaders or other heroic individuals were understood to possess a special kind of charisma or "soul stuff" that was the underlying cause of victory or defeat in warfare and the flourishing of society. Taking an enemy's head would transfer their charisma to the taker.[29]

Headhunting was a gendered activity. For men, taking heads and other displays of bravery were vital means to gain honor. In some societies, men wore tattoos to display the number of kills, and heads might be displayed or buried with the bodies of elite fighters who took them.[30] Failure could result in social stigma and shame, which could only be redeemed by success in battle. Honor and shame affected a person's prestige; in some communities, men could not marry before taking heads. Women's ritual role involved shaming men (or threatening to shame men) who did not display adequate bravery or promised glory in battle.[31]

Thus bravery in battle and the taking of heads were vital markers of masculine efficacy, while women reinforced the social stigma against men who failed to meet this standard.

Raiding, however, was not always or solely about taking heads, although it nearly always conferred prestige on combatants. Conflicts established political control, settled insults, and gained for victors plunder or territory.[32] Such "everyday" violence seems to have been relatively common in many areas, if never less than devastating for those involved. For example, archaeologists found that 25% of the human remains at one mortuary site in Luzon had died by violence.[33]

The technological effort expended on warfare was considerable. Most adult men would possess weapons and the skills to use them. Some level of preparation for warfare or raiding appears to have been a regular part of life among headhunting groups. For example, in Luzon, boys were taught to take heads of small animals using miniature versions of the bladed weapons adults carried.[34] Yet we should be careful about evaluating how much technological development was devoted solely to the production of weapons. Some weapons used for raiding were likely the same as those used for other everyday needs—most weapons could be used as much for hunting as for warfare (figure 28).[35] Likewise, amphibious raids in coastal areas would employ ships, which were also used for nonviolent purposes.

Some widely used weapons do appear to have been purpose-built for warfare, however, especially for close-quarters fighting and ambushes. In many areas, nearly every adult male could be called to participate in war, meaning that personal weapons were ubiquitous and would have been produced by local metal- and woodworkers in most cases.[36] These included spears tipped with steel, lances, blowpipes, bows and arrows, and the *kris*. The kris was a dagger crafted by elite artisans with special ritual and technical authority. It is most associated with the Malay world but was used in many parts of Southeast Asia.[37] Blowpipes were made from bamboo when available or laboriously bored out of suitable local wood. The ritual manufacture of blowpipes required considerable skill; for example, the Bugis monarch in south Sulawesi employed a royal blowpipe maker.[38]

Figure 28. Head ax from northern Luzon Highlands, used as both weapon and a tool (e.g., as an adze).
By Lorenz Lasco via Wikimedia Commons. CC BY-SA 4.0, https://creativecommons.org /licenses/by-sa/4.0/.

The favored poison used for arrows, the latex from the trees of the species *Antiaris toxicaria*, was extraordinarily lethal and required great care in preparation and use. European fighters were rightly terrified of these weapons, and European scholars searched fruitlessly for an antidote.[39] Over time, with the spread of firearms, blowpipes fell out of general use, although they are still made and used for hunting in some parts of Southeast Asia (figure 29).[40]

Long before the early modern period, weapons of war or pieces of weapons (like steel spear points) could be obtained through trade and local manufacture; the kris is a ready example of a weapon widely circulated around the region. Thus obtaining new weapons such as guns and cannons via trade would not have represented a major shift in practice, although the expense of these items was considerably higher than swords or kris.

Weapons conveyed status and spiritual strength—charisma—in many

Figure 29. A man of the Kenyah people in Sarawak (on the island of Borneo) making darts for a blowpipe.
Courtesy Wellcome Trust. Public domain.

Southeast Asian societies, as much as they did preparedness to fight. More than mere symbols of a warrior's charisma, they were sometimes understood to possess charisma of their own. The kris offers a useful perspective on the relationship between weapons, status, and spirituality in the Malay world. All adult men in the Malay world carried the kris, making it as ubiquitous as the spear. Yet the kris was far from socially ordinary in the way that spears were. Foreign observers in the period stressed that no adult male would be seen in public without a kris (many women carried them as well), and there were strict rules about how kris were displayed in the presence of royalty. They were both weapons and marks of social status, lineage, and kinship. Most men on Java, for example, had three kris—one from his ancestors, one personal weapon, and one from his father-in-law.[41]

Kris manufacture required considerable skill as the dagger blades were repeatedly folded and refolded (usually at least 64 times) to obtain their characteristic qualities, including the *pamor*, or pattern of the steel in

Figure 30. Kris dagger with sheath.
By Crisco 1492 via Wikimedia Commons. CC BY-SA 3.0, https://creativecommons.org/licenses/by-sa /3.0/.

the blade (figure 30). Blades intended for royalty or elite fighters might be subjected to rituals to enhance their spiritual efficacy. Such daggers would commonly be carried in gold or jeweled sheaths, reflecting the owner's prestige and spiritual authority (figure 31). Elite warriors in the Malay region marked their status by carrying both swords and kris.[42] Thus the charisma and spiritual power of leaders and warriors, thought to be foundational to the success of warfare, was expressed and demonstrated through the form of the kris and its use in combat.[43]

Because warfare was understood as a test of spiritual strength as much as a pragmatic activity focused on gain, soldiers and their weapons operated within a wider spiritual economy. Where elite soldiers existed, they were usually members of the aristocracy or specially trained to serve royalty (figure 32). On Bali, for example, such fighters were understood to possess more charisma than the ordinary run of soldiers and superior training or dedication. Their exceptional charisma made them better fighters and engendered the trust of regular soldiers who looked to elites for inspiration and courage.[44]

As mentioned earlier, their weapons could have charisma of their

Figure 31. Postcard image of a Balinese nobleman. Note the decorated sheath for his kris over his shoulder.
Courtesy Leiden University Digital Collections. Shelfmark KITLV 1400138.
CC BY-SA 4.0, https://creativecommons.org/licenses/by-sa/4.0/.

own. In Bugis mythology, for example, the sacred blowpipe (I Buqle) was used by the Bajeng people to defeat hundreds of warriors in the kingdom of Goa until Goa obtained the blowpipe through a betrayal. From that time forward, the blowpipe protected the people of Goa, transferring its spiritual efficacy to its new holders.[45] Guns, blowpipes, swords, bows and arrows, and kris could all be understood to carry magical powers that transferred to their holders under the right circumstances. Even when technologically out of date, certain guns continued to be held by Southeast Asian kings because of this spiritual potency.[46]

Figure 32. A portrait of an Ambonese soldier. Note the fine
garments, and weapons, as well as the soldier with the
severed head in the lower left.
In François Valentijn, *Oud En Nieuw Oost-Indien*, v. 2. Public domain.

Technological preparations for warfare went beyond the realm of men
and their weapons and into the everyday lives of women as well. Women
were responsible for creating objects essential for combat, including
banners, talismans, and charms of invulnerability. Such technologies
were as vital to the spiritual economy of warfare as were weapons. For
example, in the Malay world, invulnerability charms might be made

from personal or bodily materials, such as the hair of a warrior's sisters, sometimes mixed with menstrual blood.[47] Banners smeared with blood in the Bugis tradition were so important to the maintenance of spiritual strength that defending the banner was vital to success in battle. Women wove these banners and sometimes provided the blood, which gave them power. Women wove other ritual cloth also, including, for example, the specially patterned cloth the Iban people of Borneo used to receive the severed heads captured by warriors. Doing so involved learning a specific design used solely for this purpose.[48]

The spiritual needs of warfare affected mundane forms of daily technological production. Among the Bugis, no one would work in the fields while preparing for war, and in the Philippines, women's work in particular was expected to stop entirely during battle.[49] In Sumba (an island in the present-day nation of Indonesia), women were expected to stay indoors and refrain from sewing and weaving during times of war. In other places, it might be just one or two women who observed such taboos on behalf of a village.[50] Thus the technological life of a community was shaped by the spiritual dimension of warfare, even among those who did not fight.

In practical ways, too, warfare could have significant repercussions for the technological organization of communities. Raiding to obtain captives, either to put to work or to trade as slaves, was widespread. For example, in the Philippines, supra-village chiefs extended their political prestige and their ability to obtain valuable products for trade through raiding. Frequent raids provided captives, some of whom might be sold to pay for weapons, while others would be put to work collecting the valuable forest and sea products traded to the Chinese for tea, porcelain, and other goods.[51] As demand for these products skyrocketed starting in the 1500s, so, too, did violent slave-raiding. Such chiefs might employ well-armed mercenaries like the Iranun (an ethnic group from the southern Philippines and Malaysia) to fight for them, granting them captives as part of the payment. These raids were so devastating and violent that even 100 years after the last attacks, a deep-seated fear of raiding was still evident among the people of the area. Local villages protected themselves as much as possible, sometimes by creating escape routes, sometimes by building fortifications. Yet over the course of the early modern period, hundreds of thousands of people were sold into slavery, producing a demographic collapse in the region where the Iranun operated.[52] In less extreme cases, agricultural production was itself so valuable that warfare simply settled who agriculturists produced for, rather than upsetting the entire system of production. For example, in Burma and Siam, captives often did agricultural work in royally held fields, thus enhancing the victor's power.[53]

The expanded use of guns in the 1500s was tightly tied to this expansion of the slave trade. As European, Turkish, and Chinese firearms became more available and more desirable even for relatively small-scale disputes, Southeast Asians looking to make significant purchases often traded enslaved people, one of the few "commodities" whose value was sufficient to accommodate an exchange.[54] Thus the slave trade, although long established, expanded as the arms trade became more valuable. Firearms were used to gain more captives, and captives traded to acquire more guns.

Conclusion

Small-scale raids, village-to-village fighting, and supra-village warfare absorbed everyday technological attention from all members of society. With most adult males carrying weapons, local and regional craftspeople expended energy making arms, while others increasingly sought difficult-to-manufacture firearms through trade or conquest. Local production of charms, amulets, banners, and other ritual objects occupied even noncombatants, contributing to the spiritual efficacy of warfare. In this way, warfare permeated the technological life of even small or relatively isolated societies.

CHAPTER 7

Scaling Up Warfare

Early Modern Technology and the Warring State

WARFARE ON BEHALF OF large polities required attention to the acquisition and production of technologies of war at a larger scale than what we have so far explored. In most cases, the ultimate motivations or aims of warfare stayed the same: to consolidate or extend political control over people, territory, and resources and assert the definitive spiritual prestige that would justify that authority. However, the ways that large Southeast Asian states achieved those ends were diverse, including how they went about scaling up their armies and modifying techniques of warfare to support regional or state-level conflict.[1] Starkly different modes of warfare, from small-scale intervillage headhunting to large-scale battles for imperial conquest, existed side by side throughout the period. There was no simple evolutionary relationship between the two, nor any singular model or theory of warfare.[2] This chapter explores the technological work involved in scaling up warfare and how these projects of expansion affected wider sociotechnical order, including the effects on the lives of ordinary people far from the centers of power.

Strategies for Scaling Up

One of the most straightforward approaches to scaling up conflicts was simply to increase the number of combatants, leaving other practices, such as modes of provisioning and the overall organization of forces, the same. For example, on the island of Ambon, village warfare expanded primarily by creating alliances of villages, who then assembled a *hongi*, a fleet of ships from multiple villages. The vessels that made up the hongi were the same *kora-kora* vessels used for village-level fighting, crewed by the same people, including a leader from each village in each ship.

It is less clear whether the hongi as a whole had a separate leader or if individual vessels simply agreed to the aims of the conflict and acted on their own recognizance. Their style of raiding, however, including taking heads, stayed much the same as any village-level ship would have practiced. Fighting at all scales increasingly involved forms of firearms by the seventeenth century, although firearms may not have been as pragmatically effective as traditional swords and spears. Problems with maintenance and the time required to reload during battle may in some instances have made them more frightening than deadly. Even as ambitions for territory and people grew, warfare in Ambon changed only modestly.[3] The VOC itself adopted a hongi for coastal fighting when they attempted to exert control over the region.[4]

Similarly, on Bali, the near-constant state of warfare with kingdoms of Lombok in the seventeenth and eighteenth centuries resulted in larger and larger armies. Yet, provisioning, weapons, and tactics were similar to those deployed in smaller-scale fighting. Ordinary fighters continued to supply their own personal weapons like lances and javelins just as they had in smaller conflicts.[5] Elite soldiers used firearms, blowguns, and swords. Both types of fighters were deployed in fierce frontal assaults (called *amok* attacks) that varied little from earlier, smaller conflicts.

Visual evidence from Burma in a later period (the nineteenth century) likewise suggests that peasant conscripts, who made up the vast majority of fighters, went to war with nothing more than their own weapons. Although some activities, like the taking of heads, were frowned on by officers, peasant fighting was nevertheless much the same in large-scale battles as it was in village-level conflict.[6] It seems that from the perspective of the conscript, the experience of fighting in large-scale wars might not have been all that different from village-level fighting.

However, when scaling up warfare involved increasing numbers of professional or elite soldiers, the situation was different. When states sought to increase the numbers of elite fighters, they often pursued more thoroughgoing changes to training, arming, and provisioning. A good example is the kingdom of Đại Việt, where significant institutional changes to armies are visible from no later than the thirteenth century. Officers (normally drawn from elite families) underwent training that tested strength, skill, and knowledge. Such training is documented in treatises on military practice authored by the imperial Vietnamese prince Trần Hưng Đạo (1228–1300), who seems himself to have drawn in part on Chinese military manuals.[7]

Professional or elite soldiers in whatever part of the region were usually specialists. They were more likely to participate in some training because they were more likely to carry weapons that required training for effective use. For example, the Bugis and Makassarese peoples (both located in the southern part of Sulawesi) employed a professional class of soldiers who were highly trained with weapons like swords and kris; they also trained to fight from horseback, and later with firearms. Such soldiers were referred to as "fighting cocks" and were seen as especially well prepared for combat both physically and spiritually.[8] Although documentation about the training of Balinese soldiers is absent, it is clear that the VOC were strongly impressed by their fighting skills, suggesting that elite fighters were well trained.[9] The presence of trained soldiers in the early modern period was not new; many societies had elite fighters with special skills they honed for battle. The question is what role these professional soldiers played in fighting and how efforts to support their needs might have expanded (figure 33).

The use of mercenary organizations is a case in point. Mercenaries—viewed by Europeans as pirates—offered a straightforward, if expensive, way to increase the scale of conflicts for ambitious leaders who did not wish to (or could not) raise or provision armies of their own. In the Sulu

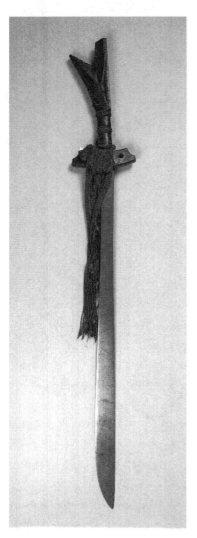

Figure 33. Fine swords like this kampilan from the Philippines required considerable skill and training.

Courtesy Metropolitan Museum of Art via Wikimedia Commons. CC BY-SA 1.0, https://creativecommons .org/licenses/by-sa/1.0/.

region, supra-village chiefs hired the mercenary Iranun people (described in chapter 6) to raid villages and take captives on the chief's behalf. The Iranun grew to be one of the most technologically sophisticated fighting forces in the region. By the late eighteenth century, the Iranun could attack with as many as 40–50 ships of up to 130 feet in length, armed with the best gunnery available and crewed by 2,500–3,000 men. Funded by the slave trade, they presented a significant threat to settlements of any size and routinely damaged Dutch and British trade.[10] The Iranun undoubtedly bought ships and guns, but other aspects of their training and provisioning are less clear in the sources. It seems likely that they employed artisans of their own and invested considerable efforts to maintain their military advantages, both in training and technology. Further study would undoubtedly provide more insight into how technological skills and abilities were organized on behalf of this mercenary warfare. However managed, Philippine leaders and others scaled up warfare by deploying this preexisting "war machine." Therefore, those who hired the Iranun spared themselves both the effort of training and the constant disruptions to economic activities that fed the all-important China trade that a war fought with conscripts would entail.

A more comprehensive military transformation than any of these examples comes from Vietnam. Although not the only large imperial state (others include Burma and Siam), and by no means representative of all imperial states, Vietnam's experiences are exceptionally richly documented.[11] As mentioned earlier, there is evidence of formalized officer training in Vietnam as early as 1200. However, the aims of warfare at that time would have differed little from what had come before. Large-scale raids focused on captives and plunder settled disputes over honor or political hierarchy without threatening the overall balance of power between regions. Motivations started to change, however, with the invasion of Đại Việt by China's Ming dynasty in 1406. The Ming invaders exposed Vietnamese to both gunnery and a bureaucratic and imperial logic of governance that extended to military affairs. Subsequent Vietnamese leaders embraced both new weapons and a new logic of warfare that resulted in a major change to the practices and technological intensity of warfare in the region.[12]

Guns were critical to the success of the Ming invasion and the 22-year occupation of the region. The Vietnamese gained detailed knowledge about guns and gun-making because Vietnamese men were trained to serve in the imperial Chinese army during the occupation.[13] In 1428, the nascent Lê dynasty, under the leadership of Lê Lợi, was able to use this knowledge to acquire and use guns in a rebellion that threw off Ming rule. Their expertise with firearms and a reliable supply of guns and gunpowder became a decisive advantage for Đại Việt regionally. Despite throwing off Ming rule, they adopted the techniques and imperial viewpoints of their former masters.[14]

The Lê dynasty brought new ambitions and highly skilled use of firearms to their conflicts with an old enemy, the Kingdom of Champa. When the Vietnamese decisively defeated Champa, which had not adopted guns, in a brief but devastating battle, the advantages of firepower and skill were evident.[15] Yet the full story is hardly one of technology solely driving the outcome. Guns helped the Vietnamese gain victory, but their changing political aims and goals strongly shaped the balance of power in the region. The Vietnamese incorporated much of Champa into their own empire and forced cultural transformations on the conquered peoples in ways unprecedented in the history of their conflicts. Exposure to Chinese imperial philosophies, which stressed the need for new territories to conform to their conquerors' modes of social and political organization, strongly influenced Lê Lợi and subsequent Vietnamese emperors. The logic of warfare had changed, and therefore the results of warfare for the people of Champa involved deep changes in their ways of life. Guns alone did not determine this outcome.

Even if we focus mainly on the technologies of warfare, guns were not the only technologies the Vietnamese deployed to achieve victory. Information technology, including extensive geographical knowledge, mediated by technologies such as maps and route marches, now used for imperial ends rather than merely for raiding, determined the best placement of guns in their decisive battles against Champa.[16] Although the Chams were unable to offer much resistance, other Tai peoples did possess guns and presented a far more difficult challenge. Vietnamese victories against these opponents depended on the Vietnamese military organization's increasingly sophisticated training and record-keeping.[17] Other forms of military modernization, such as preventing generals from holding land in their own right and drawing on the central authority of the emperor to levy conscripts, were equally important for creating a highly effective and focused warfare state. Records suggest that armies were very large; the troops who fought the Chams were reputed to be 300,000 strong.[18] Even if most conscripts supplied their own weapons, records of officer development and the acquisition of arms and ships still suggest formidable and increasingly centralized sociotechnical effort.

However, sustaining focus on preparing for warfare across generations did present a significant challenge, especially when the state fragmented into civil war. Although Vietnamese armies put considerable effort into acquiring military advantage and upgrading their manufacturing capabilities during warfare, they failed to maintain consistent technological preparedness. For example, in the seventeenth century, Trinh leaders in the north, who were engaged in a civil war with the Nguyen in the south, invited the VOC to set up a gun foundry. Their Nguyen rivals worked with the Portuguese for the same purpose.[19] The Nguyen also spent time developing skills in casting to reduce their dependence on the Portuguese. Over time, the Nguyen boasted more than 1,200 bronze cannons, which were critical for defeating the numerically superior Trinh.[20] Yet by the time of the Tây Sơn rebellion in the eighteenth century, skills with guns, military training, and gun manufacture had dropped off, leaving the Nguyen leaders scrambling to bring their army and navy up to date. They again tapped foreign military advisors, this time the French, to improve their fortifications and develop saltpeter manufacturing. They also obtained a Chinese translation of a European military manual on artillery and tactics.[21] Their rivals, the Tây Sơn, also looked elsewhere, in their case Bengal, to enlist aid in training their fighters in the use of modern naval guns.[22]

Vietnamese leaders built their naval capabilities by drawing on their own considerable local expertise in shipbuilding and the relevant and up-to-date experiences of others. Notably, Vietnamese ships were vital not just for warfare but also as a commodity to trade for other necessary war materials, including guns, saltpeter, and iron.[23] Thus trade, the general circulation of textual knowledge about weapons and tactics, diplomatic contacts, and local manufacturing abilities were all part of the dynamic technological infrastructures the Vietnamese used when engaging in conflict on a large scale. Vietnam's experience may not be simplistically representative of other major powers in the region. Still, it offers insight into the challenges of maintaining the capacity for large-scale warfare when arms and tactics were changing rapidly.

Reconfiguring Spaces and Circulating People: Technological Consequences of Conflict

As was true with small-scale conflict, the story of technology and warfare is not just about the development or deployment of technologies to fight or support battle but also about the ways that warfare reinforced or reconfigured the sociotechnical foundations of Southeast Asian societies. Larger-scale conflicts could result in significant changes to the spatial and demographic organization of these societies and their sociotechnical organization of production, although these consequences have in general not yet been deeply researched. For example, when raiding in Ambon and other islands like Seram and Tidore became more intensive in the sixteenth century, villages moved out of the vulnerable coastal regions and into interior or upland areas that were more defensible. This shift made trade more difficult, suggesting at least the possibility that the focus of local production may have shifted more toward agriculture to compensate. The growing pressure of warfare on the Balinese population played a significant role in the political fragmentation of Bali in the seventeenth and eighteenth centuries.[24] What the technological results of such fragmentation may have been remains to be studied.

At stake in many conflicts was access to agricultural land and laborers. On Java in the seventeenth century, soldiers of the Kingdom of Mataram repeatedly burned the rice fields that fed their enemies in the Kingdom of Surabaya, cut off access to timber supplies, and dammed the river that provided water to the city. Mataram depopulated conquered cities and villages, moving captives to areas under Mataram's control. These tactics successfully asserted Mataram's dominance at the cost of the agricultural base. When an El Niño weather pattern occurred in the early seventeenth century, severe food shortages and deaths by illness became widespread. The VOC, which had hoped to trade in coastal areas for surplus rice, were turned away, setting in motion conflict, which further ravaged the area.[25]

What is clear from these brief examples is that contests over territory were also always contests over labor. People with special technological skills were particularly desirable. Looking closely at habits of captive-taking and trade demonstrates how dramatically these practices were intertwined with technological dynamism. Although slaving was nothing new in this period, it seems to have intensified, with substantial consequences for the everyday lives of people in the region. As already mentioned, the presence and increasing value of guns meant that trade in guns and captives were constantly intertwined in ways that tended to reinforce conflict and the involuntary migrations of captives.[26]

When captives were held rather than traded, they were often put to work in ways that would enhance the captors' wealth and prestige in the changing trade environment of the early modern period. As already mentioned, Philippine chiefs used captives to collect the forest products demanded by Chinese merchants. The Konbaung dynasty in Burma put captive laborers entirely under royal control to increase their personal wealth. By taking so many captives, the Konbaung simultaneously depopulated the countryside of their primary enemy, the Kingdom of Ayudha. The loss of population and the labor they provided contributed significantly to the fall of Ayudhya in 1767.[27]

Technologically speaking, captives significantly augmented royal power. The size and ambition of state technological projects were correlated with the acquisition of enslaved people to work on these projects. In Mataram, captives from Surabaya were resettled in interior rice lands under the direct control of the sultan, increasing the power of the sultan and the agrarian base of the empire simultaneously. In Burma, captives were divided into groups of skilled artisans, agricultural laborers, and unskilled laborers. Skilled artisans became the property of the king and were settled in villages according to their expertise. Agricultural workers were put to work laboring in royal rice fields, and unskilled laborers were put to work on public works projects.[28] In eighteenth-century Burma, 35% of royal artisans were captives.[29] Although deprived of their basic freedoms and kinship networks, skilled royal artisans in Burma were held in high esteem and paid no taxes. Especially prized were those skilled workers who could support the warfare state—for example, gunpowder production.[30]

In the mainland coastal kingdom of Arakan in the seventeenth century, the Arakanese captured Portuguese freebooters and put them to work as slavers. The people the Portuguese captured were made to grow rice that the Arakanese leadership traded with the Dutch. The Dutch used this rice to support their fighting in Ambon.[31] King Taksin, who led the successor state (located in present-day Thailand) after the fall of the kingdom of Ayudhya, and King Rama I of Siam (1782–1809) used similar tactics. They took captives whenever possible and tattooed them to make escape difficult. These captives were vital to political centralization and the economic recovery after the fall of Ayudhya. Under Taksin, the country had trouble supporting even its subsistence needs for rice, much less providing the surpluses that kept Ayudhya afloat. Later, in the Kingdom of Siam, captives were often put to work producing rice and gathering forest products, both of which had become increasingly important to the economy. By the nineteenth century, many captives were moved to royal sugar estates, a vital export crop and source of wealth for Siam's royalty.[32]

The accounts available of forced migration as a consequence of warfare only hint at how these migrations may have affected circulations and development of artisanal knowledge, the growth of estate-style agriculture, and technological choices about day-to-day subsistence in the face of military threat. More historical attention to all of these would give us both a better understanding of the technological dimensions of warfare and its integration into wider patterns of social change.

Scale, Technology, and Comparative Advantage in Warfare

The question of comparative advantage has loomed large in the history of Southeast Asian warfare, particularly concerning the growing use of firearms. Did possession of firearms or other sophisticated military technologies hand an unbeatable advantage to larger Southeast Asian states? Did Europeans have such a strong advantage that it accounts for later European domination in the region? To reduce nineteenth-century European imperialism to a victory of arms alone is a bad misreading of a complex history. Yet the subject of comparative advantage still warrants attention. Recent scholarship emphasizes the historical contingencies that made firearms or particular designs of sailing ships more or less decisive in early modern engagements, exploring why, despite their real efficacy, new military technologies rarely offered the clear and sustained advantage that some earlier histories have claimed.

Firearms

The acquisition and use of guns tended to reinforce centralized state control. Mainland states, which fared quite well against European attempts at encroachment until the nineteenth century, saw real growth and political stabilization as they incorporated firearms into their military in this period.[33] Yet close attention to individual states undermines any sense that firearms alone are responsible.[34] Certainly, new firearms technologies contributed meaningfully to success in some conflicts, which in turn resulted in the growth of empire, as Vietnam's victory over Champa shows.[35] Yet Vietnam's decision to enact a comprehensive political and cultural transformation of Champa in their own image relied not just on the effectiveness of their guns but on the adoption of Chinese models of empire and theories of warfare. How that changing worldview was affected by firearms is unclear.

There is good evidence that states of all sizes made serious efforts to acquire guns, from big siege weapons to matchlocks, and invest time in training elite soldiers, demonstrating the value that historical actors ascribed to firearms for maintaining political control (figures 34 and 35).[36] Small polities like Lan Na (located in what is now northern Thailand) and Shan states (in Myanmar) acquired firearms and firearm skills from the region of Yunnan long before European weapons were available. They were able to better defend themselves and maintain their sovereignty in part because of their capabilities with firearms, a lesson neither Champa nor Ayudhya fully grasped.[37]

Did firearms, therefore, confer a lasting advantage to larger states or set the stage for large-scale European domination? The answer to these questions is less clear. The ease with which large and small firearms were acquired and incorporated into existing modes of warfare meant that, although more than a trivial undertaking, arming themselves with

SEAL OF RAJAH BOOJONG OF TROUMON, SUMATRA, AFFIXED TO A LETTER SENT TO CAPTAIN
JAMES D. GILLIS OF THE SHIP "BORNEO", INVITING HIM TO VISIT HIS PORT FOR PEPPER
AND EXPRESSING A DESIRE TO PURCHASE A PAIR OF GUNS.
From the original letter in possession of the Essex Institute.

Figure 34. Letter to an American ship captain from a Sumatran leader offering
to trade pepper for guns.
In George Granville Putnam, *Salem Vessels and Their Voyages.* Public domain.

these new weapons was something that even small states could do. Southeast Asian states did not have to transform their institutions of warfare
to use them. Guns were as easily embedded in the systems of meaning
attached to warfare as in fighting itself. Large guns were regarded as
holders of charisma and thus both symbolized and reinforced the potency of rulers. Guns required training and maintenance, a sticking
point for smaller polities, but adding firearms to their arsenals seems
not to have been disruptive or controversial. They were used in both
small and large conflicts.

The use of firearms did not necessarily result in changes in tactics
in warfare, although they might provide more dramatic results. Guns
were used straightforwardly in both siege and defense. They could be

Figure 35. This fifteenth-century cannon was cast in Spain and made it to Southeast Asia in the hands of the Portuguese. By the eighteenth century, it had become part of the armaments controlled by the government in Arakan (in southern Myanmar, now called Rakhine). In the nineteenth century, it was turned over to the British. It currently resides in Edinburgh.
By Ad Meskens via Wikimedia Commons. CC BY-SA, https://creativecommons.org/licenses /by-sa/4.0/.

mounted on ships of either European or Southeast Asian design with more or less success.[38] Guns gave extra authority to sieges, and combatants with guns held clear advantages over opponents armed with traditional weapons or poorer firearms in some conflicts, although early firearms were not always as accurate or as deadly as traditional weapons. The Portuguese defeat of Malacca in 1511, for example, owed much to the superior ship-mounted guns of the Europeans. However, such advantages might not last long. By 1568, the Sultan of Aceh used similar tactics to place Malacca under siege, weakening the Portuguese imperial presence.

Moreover, Europeans continued to find traditional weapons, especially blowpipes, fearsome; blowpipes could be more accurately aimed than matchlocks and might have been more deadly in some conflicts.[39] The spread of firearms did not necessarily change the range of strategies and tactics employed by combatants. Europeans and Southeast Asians alike engaged in everything from large-scale battles to small-scale raid-

Figure 36. A nineteenth-century Indonesian matchlock.
By Verosaurus via Wikimedia Commons. CC BY-SA 1.0, https://creativecommons.org/licenses/by-sa
/1.0/.

ing.[40] From this perspective, guns may have shifted advantage in individual battles or wars, but they did not broadly transform warfare.[41]

One reason guns provided little lasting advantage to any party was because they were relatively easy to acquire. Given the high value of Southeast Asian goods and the preexisting trade in enslaved people, the most common ways for Southeast Asian polities to gain firearms were to trade for them or be gifted them by allies. The marketplace soon provided many options. For example, Turkish suppliers provided large siege guns, and the Portuguese supplied reliable matchlocks.[42] Some local manufacturing emerged as well, especially of smaller, more easily manufactured guns, like matchlocks and snaphaunces, and gunpowder.[43] For example, the kingdoms of Siam, Aceh (on Sumatra), Burma, and Vietnam engaged in gun manufacture of varying levels of quality to make themselves less vulnerable to outside providers (figure 36). The Javanese Sultanate of Mataram (1587–1755) made rifles, as the Balinese sent smiths to Mataram along with diplomatic envoys to acquire these skills.[44] Preexisting expertise in metalworking accounts for the ease with which these areas established gun manufacture. For example, the stateless Wa people of the mainland, who were long established as skilled metalworkers, became known for their ability to repair guns.[45] The capture of gunsmiths in war and the diplomatic exchange of knowledge may have contributed to the transmission of reliable expertise.[46]

Maintenance of guns and husbanding of scarce resources were critical to maintaining a working body of firearms. In these areas, the centralized control and superior resources of larger states might have been a real advantage. For example, the Ambonese used guns against the VOC in battle, but they appeared to have trouble maintaining them, giving the VOC an edge in conflicts.[47] The Javanese, by contrast, with poor access to supplies of gunpowder, created bamboo cartridges that not only kept powder dry but contained precisely the correct amount needed to avoid waste.[48] Any army that failed to prioritize maintenance of their firearms, powder, and shot soon found themselves with no advantage apart from the spiritual benefit that guns were understood to provide. More research into how cultures of artisanship, manufacture, and repair were brought to bear on warfare and, conversely, how preparation for war may have benefited the growth of skills in non-war-related activities would be beneficial for understanding more clearly the broader consequences of firearms and warfare on Southeast Asian society.[49]

Sailing Ships

The technological superiority of naval technology has also commanded significant attention. The Portuguese defeat of Malacca in 1511, as mentioned earlier, is frequently attributed to the superior naval capacities of Portuguese ships that could mount large cannons. The strength of the bombardment and the comparative superiority of Portuguese artillery played a significant role in that defeat, although disaffection with the current sultan by some merchants has also been given some blame.[50] In conflicts on the deep ocean, European ships were almost certainly superior to most of the Asian ships they might have encountered, at least at first. They were large; able to carry crews, marines, and cargo to match; and designed specially to carry heavy guns. They were faster when sailing against the wind, countering Chinese junks' superior speed when sailing with the wind.[51] The idea that warfare was an essential partner to trade played out clearly in European ship designs.

Asian leaders took note, however. In the 1630s, the Chinese pirate Zheng Zhilong took cues from the strong and highly defensible VOC ships when building his fleet, including reinforced decks to hold 36 large guns and European-style gun ports.[52] Learning from European ships was not difficult in terms of sheer knowledge but could favor large states with good resources. In Southeast Asia, the skilled naval architects working for Nguyễn Ánh in the late eighteenth century bought a European vessel at his command to dismantle it and understand its design. Smaller actors could not afford such research. Although never a large part of Nguyen's impressive navy, it is clear that historical actors found both the technologies themselves and the technical knowledge embedded in them potentially valuable.[53]

Even when Southeast Asian navies could not or did not adopt European designs, there were ways to counter European advantages in battle. Surprise attacks and shifting the field of conflict away from the deep ocean were two strategies used frequently. For example, Burmese river-going vessels—oared ships with shallow drafts—could be armed both fore and aft and quickly change the direction they moved. Such ships could outfox the large ships that tried to follow them up the river.[54] Similarly, Vietnamese crafts were more flexible than their European counterparts in operating better in the treacherous conditions of the lower Mekong.[55] These ships could mount guns, although retrofitting ships not designed for guns was no easy task, as guns will affect the distribution of weight and thus how they handle in the water. Making sure ships with added guns behaved as desired would have required considerable skill.[56]

Although the notion of an unbeatable European ship is for all of these reasons unsustained by historical evidence, it is equally true that there were limits to the advantages provided by local ships and local knowledge. Most ships could be quickly outclassed if used outside their intended environment. Like Southeast Asians, Europeans considered the full range of ship options available to them when fighting—flexibility was a widely shared naval trait. For example, Europeans used Mediterranean-style galleys against Burmese riverboats, and the VOC happily employed Ambonese *kora-kora* while they worked to establish control over that island.[57] The main advantage Southeast Asians had lay in their detailed knowledge of the environments in which conflicts took place, a considerable advantage, but one that could diminish if Europeans fielded experienced commanders.

When studying conflict in early modern Southeast Asia, historians have long stressed that shifting strategies of political alliance between diverse Southeast Asian, Chinese, and European actors undercuts simplistic "East vs. West" framings of history. Yet there remains a tendency to see technologies of warfare in those terms. The long-term background of exchange in objects, knowledge, and technique between Europeans and Southeast Asians shows that analyzing comparative technological advantage in terms of "Eastern vs. Western" technology obscures more than it reveals. Not only is putatively "Western" technology strongly shaped by cosmopolitan influences, and thus hardly "Western" in any simple way, the constant and easy flow of technologies of warfare around the region tended to level the playing field, at least among those who could muster the resources adequate to their ambitions. Over the long term, no players in the region, whether Southeast Asian, Chinese, or European, gained an unbeatable, long-term advantage from either firearms or ships.

Conclusion

Throughout the early modern period, warfare occupied a considerable amount of technological attention in Southeast Asia, providing and maintaining arms and other ritual and practical materials of war, and training elite warriors to fight on both small and large scales. Although warfare did increase in this period, humble conscripts, used to coming when called on with their own weapons, might not have noticed much of a break with the past. Although some fighters would find themselves armed with matchlocks in addition to spears, swords, or kris, the ways they were organized to fight and the reasons they fought may not have differed much from earlier times. Recent scholarship has decisively challenged the idea that European ships or firearms comprehensively revolutionized warfare and offered an unbeatable advantage to Europeans in conflicts.

The historical questions we can ask about military technology in Southeast Asia go well beyond questions about technological determinism or firearms revolutions. One approach that rejects the Eurocentrism of earlier scholarship is to pay attention to how Southeast Asians deployed technology when they scaled up warfare during the early modern period. Some polities scaled up by adding more people to warfare, others by adding more powerful weapons, and most by doing some of both. Many worked to arm their own fighters, while a few called on mercenaries, who, over time, developed into a technologically formidable regional force. Devoting more resources (including people) to warfare sometimes had devastating consequences for sociotechnical ways of life, especially for captives whose expertise was highly desirable. Farmers forced onto new lands, artisans deprived of their family networks and freedoms and made to produce goods for a royal master, and ordinary people put to work gathering valuable forest products all paid a high personal cost for warfare. War disrupted established ways of life, justified intensified exploitation of land, people, and resources, and unwittingly circulated technical knowledge in its wake. Large-scale fighting had serious implications, even for noncombatants.

Even if current research does not yet allow them to be completely mapped out, the implications of these migrations underscore Tansen Sen's point that conflict as much as cooperation shaped the character of regional integration in Asia.[58] If it is true that warfare accelerated the circulation of technical knowledge on military matters, agriculture, and artisanal production, then the early modern period saw increasing regional engagements designed specifically to achieve technological ends.

CHAPTER 8

Technology, Empire, and Nationalism

T HE NINETEENTH AND EARLY TWENTIETH centuries saw dramatic shifts in Southeast Asian life, much of it connected to the spread of European colonialism. Extractive enterprises such as plantation agriculture, forestry, and mining expanded to feed global markets, and environments changed in turn. Population densities increased, transforming Southeast Asia from labor shortage to labor surplus.[1] Widespread colonialism transformed social orders and political relationships even in places where colonial rule never took hold, such as in Siam, and spawned new forms of political organization and resistance where it did. Southeast Asian peoples, on average, saw their standards of living decline by the end of the nineteenth century, although decreased warfare may have helped some see improvements in their quality of life. Incoming Western cultures, including technological cultures, both attracted and repelled. What role did technology play in this history of change? Two themes stand out in existing literature, the consequences of Euro-American industrialization for the increasing integration of Southeast Asia into non-Asian markets and the role of incoming Western technology in transforming agriculture, manufacturing, labor, and mobility in Southeast Asian life. Historians interested in technology have given special attention to the role of technology and technoscientific projects in the wider politics of empire in the exploitation of Southeast Asian peoples and environments.[2]

These themes certainly matter, yet their formulation betrays foundational problems in thinking about technology in modern Southeast Asia. The focus on technology introduction, or incoming technologies, can make it impossible to see the significance and the central place of preexisting Asian technologies in the dynamics of change.[3] The relationship between market demand and local technological transformation requires careful treatment to avoid technologically deterministic analyses that fail to account for the complexity of Euro-American and Asian technological contexts. Both themes tend to analytically reinforce East-West binaries, poorly representing technological life as it was lived. Although much scholarship has explored how such binaries were constructed as a way of asserting colonial power, the tendency to see some technologies as "Western" in origin or ownership reinforces a view of Southeast Asia (and other colonized places) as sites of technological deficiency in the nineteenth and twentieth centuries. Such thinking correlates Western colonial success with Western technology in ways that are misleading both technologically and politically.

An example of this is the "tools of empire" thesis.[4] Briefly, this thesis argues that Europe's industrial changes, including key technologies like steamships and railways, made colonization cheaper and more efficient and thus facilitated imperial expansion and control. It focuses on particular technologies whose European origins are clear (weapons, steamships, communication technologies) and explores the value of these technologies in warfare and the operation of colonies. The main problem of this thesis is that it overplays both the significance and the efficacy of such technologies in many colonial undertakings. Without questioning whether these technologies had value and significance, a harder look at the actual practices of colonialism calls the overall thesis into question. Some techniques of control were not technological at all, such as the use of differing legal systems based on ethnicity or other "divide-and-conquer" forms of organization, changes to laws about land use, and similar administrative practices.[5] Such techniques might render simple technologies like identity papers (hardly a Western invention) into agents of control as powerful as the most lethal gun. Moreover, technologies such as railroads may have complex outcomes, distributing advantage and disadvantage in ways this thesis fails to account for. This is not to say that we should dismiss the significance of incoming technologies from Europe or elsewhere. During no period of Southeast Asian history was the incoming flow of ideas, techniques, or technical artifacts insignificant. Rather we should understand nineteenth-century incoming technologies in much the same way as those from earlier periods: exploring how technologies are integrated into or assembled with preexisting ways of life and the local contingencies that shaped how these technologies were co-constructed with political or social power.

Rejecting a view of Southeast Asia as "insufficiently technological" because of nineteenth-century European technological change, recent scholarship explores the emergence of new forms of sociotechnical order assembled from both preexisting and incoming technologies. The resulting assemblages, much like those produced in earlier periods (as explored in chapter 2), were hybrids of new and old and coproduced with social and political life: they do not solely drive empire nor simplistically fulfill the desires of imperial actors.

The focus on hybridity and assemblage brings Southeast Asian technological practices back into nineteenth-century history. It opens up important questions about the technological transformations experienced by the region's peoples. How were different technologies assembled to constitute working systems, and what relations of interdependency between different technologies were created or maintained? What technopolitics and affordances emerged from these new sociotechnical arrangements? What meanings did Southeast Asians assign to technologies and sociotechnical change more broadly to make sense of their changing lives in this period? Hybridization and assemblage point to how technologies were embedded, made locally workable, durable (or not), and meaningful.

Agricultural Transformations

The nineteenth century saw dramatic increases in the large-scale production of agricultural commodities for export. Sugar, tobacco, rice, coffee, rubber, and other crops expanded in acreage throughout the nineteenth and early twentieth centuries, setting the stage for the significant role export-oriented agriculture still plays in the economies of most Southeast Asian countries. Much existing scholarship on these expansions has focused on the exploitation of Southeast Asian labor and the linkage between declining living standards among Southeast Asians and the spread of large-scale production.[6] Shifting attention to the technological dimension of these histories underscores important nineteenth-century continuities with long-term processes of agricultural intensification around the region, especially the role of Southeast Asian knowledge, technology, and technical skills in the process. Governments and planters employed a variety of approaches to increase the scale of production and not just the stereotypical plantation, which used wage (or enslaved) laborers in combination with modern plant breeding and cultivation techniques and favorable agrarian laws. Various foreign and Southeast Asian actors created new configurations of local and incoming technologies, crafting new systems to increase production. Four case studies of large-scale export-oriented production show the different processes of sociotechnical assemblage in play around the region and the consequences of those choices on people, technology, and the environment.

Rice in Burma

Burma became a major rice exporter to Southeast Asian, American, and European markets by the late nineteenth century. This massive expansion of production happened under British colonial rule yet shared much with earlier periods of agricultural growth; it relied entirely on preexisting technical knowledge and skill and was powered by migration. Before the British takeover in 1852, the southern part of Burma, a

large delta region of mangrove swamps and forests, was sparsely populated. The defeated Konbaung dynasty had forbidden rice export to protect their food supply. The British encouraged migration to the south,
anticipating cotton or tobacco production. Neither made ecological
sense, and farmers chose instead to grow rice. They employed methods
and tools traditionally used in the north and traditional family labor
arrangements and sharecropping.[7] They brought considerable skills to
bear as they transformed local ecologies, reclaiming land from swamps
and building embankments to protect land from flooding.[8] Colonial
officials did some road and canal improvements, but Burmese farmers
handled the bulk of infrastructural work. The most significant technical contribution from the British was the establishment of steam shipping on the Ayeyarwady River to carry migrants to the south.[9]

Burmese farmers migrated because of the opportunity to acquire land
and the more reliable rainfall than in the north. Rice exports doubled
between 1867 and 1873, reaching a half million tons. By the 1930s, Burma
exported nearly three million tons of rice grown on eight million acres
of land.[10] In the early phases of this expansion, farmers benefited significantly from the growing international rice market. But this did not last.
Over time, soil fertility, and thus yields, declined, although absolute tons
of rice for export continued to rise until most of the land had been enclosed. Toward the end of the period of growth, incoming farmers worked
marginal lands, using rice varieties such as Mayin paddy that could with-
stand extensive flooding. Efforts by the British to improve the situation
were either poorly thought out—such as the introduction of tractors on
consistently flooded land—or unworkable, like insufficient quantities of
improved rice varieties or expensive fertilizer.[11]

As agricultural debt increased, land ownership became more con-
centrated, resulting in overall declines in living standards. The problems
were visible even in the bodies of the Burmese themselves, who showed
physiological signs of malnourishment, including a decrease in average
stature, in the early twentieth century.[12] Southern Burma's expanded rice
production fed steadily growing markets using tools and skills in use
for centuries. Demographic expansion and environmental transformation made this growth possible. Yet the economic rationality that moti-

vated rice production produced a monocropped landscape that proved incapable of sustaining either the boom or economic aspirations of the migrants who powered it.

Sugar Planting on the Island of Negros

Throughout the nineteenth century, sugar production in Southeast Asia expanded. Despite competition from sugar beets grown in temperate climates, cane sugar remained an appealingly profitable commodity for local and foreign trade.[13] As was already clear in the eighteenth century, successful sugar production relied on a predictable labor supply, especially for the rapid harvest. Yet it also relied on technical skills to cultivate, extract, and refine sugar from the cane. Sugar cultivation moved to the previously uncultivated island of Negros in the province of Iloilo in the Philippines in the middle of the nineteenth century. As in Burma, migration and Southeast Asian agricultural knowledge were vital for establishing a sugar industry on Negros, as was a willingness to learn and experiment.

Sugar planting on Negros started in the late 1850s, primarily by Chinese mestizo entrepreneurs financed by foreign merchant houses.[14] Most of these aspiring *hacenderos* (planters) had seen their lucrative trade in locally produced textiles decline with the opening of the port of Iloilo to world trade in 1855. As an influx of cheap textiles cut into their profits, some shifted their attention to sugar. But with no experience in sugar cultivation or any other kind of agriculture, hacenderos had to recruit farmers from surrounding islands to work as share tenants. The new share tenants brought general agricultural skills for producing crops like rice or vegetables but no experience with sugar. It is unclear whether the new hacenderos (some of whom might have had holdings of no more than a few hectares) employed sugar experts from elsewhere to teach their new tenants.[15] It seems more likely that farmers developed specialized knowledge through experience with local environments, experiment, and observation. Improvements in yields in later decades suggest a substantial growth in local technical knowledge of cane cultivation.[16]

For share tenants, sugar plantations offered real opportunities. Under

Spanish rule, farmers were tied to their land. Moving was bureaucratically difficult and involved costly fees. Hacenderos offered inducements (sometimes facilitating illegal relocation), including paying the family's taxes and providing housing. Tenants received access to land, draught animals, equipment, and a cash advance on the harvest, earning half of the harvest minus advances and milling expenses. The possibility of better land, a lucrative crop, and a fresh start was attractive. Socially, share tenants were treated respectfully by the hacenderos who needed them; choosing to move was potentially a financial and social improvement.[17]

Changes to the business and technology of sugar production began to affect this mutually beneficial arrangement in the late nineteenth century. Globally, sugar producers saw disease, and significant yield declines, spurring experimentation with new cane varieties and fertilization regimens.[18] Milling technology became increasingly sophisticated and expensive, and buyers came to expect higher-quality sugar.[19] On Negros, share tenants disproportionally bore the burden of these expensive changes without increasing their share of the profits. In the 1880s, when a worldwide drop in sugar prices rendered many small plantations uncompetitive, planters increasingly abandoned their own mills in favor of paying central mills, owned by wealthy merchant houses, to process their cane (figure 37).

The central mills required relatively consistent cane quality, something difficult to achieve with share tenants where differences in skill, labor availability, and financial and ecological circumstances affected their results. Share tenants' profits dwindled, and their social status declined. By the 1920s, hacenderos replaced share tenants with wage laborers and skilled overseers, obscuring the role that farmers' technical skill and initiative had played in establishing the industry in the first place. The growing technical sophistication of milling technology certainly affected the ways that sugar could be made profitable here, as it did elsewhere. But the sociotechnical arrangements that increased tenants' costs without increasing their access to profits and the growing availability of what was once a scarce resource—agricultural experience—

Figure 37. The Central Azucarera Sugar Mill in Negros, the Philippines.
By Hbaileros via Wikimedia Commons. CC BY-SA 3.0, https://creativecommons.org/licenses/by-sa /3.0/.

also transformed sugar from an opportunity for Filipino farmers to simply another source of poorly paid labor.[20]

Sugar on Java

On Java, as on Negros, early nineteenth-century efforts to produce sugar relied on the skills and knowledge of local farmers and their ability to improve yields through experiential learning. Unlike Negros, which had been largely uncultivated before the introduction of sugar, the integration of sugar with other crops on Java had technological repercussions far outside sugar regions themselves. Although sugar had been grown on a small scale on Java in the early nineteenth century, its first significant expansion came under the Dutch Cultivation System, implemented in the 1830s.[21] Under the Cultivation System, villages produced export crops, including sugar, as a form of tax. The Dutch government then sold the crops on the European market. The Cultivation System did not involve technical intervention, thus relying on indigenous technical knowledge and skill.[22] Those who grew sugar did not become specialists. Instead, they moved between sugar and rice or vegetable production as required by village heads who distributed the compulsory labor.

The Cultivation System reconfigured established sociotechnical arrangements, requiring extra labor from farm families but sometimes also providing higher incomes. Women, in particular, saw more work in cane fields, a temporary arrangement that became permanent. They spent less time weaving cloth or growing cotton, both sources of independent income, shifting power relations within families.[23] Farmers increasingly double-cropped rice to compensate for reduced rice acreage, using faster-growing but less profitable varieties.

However, sugar production also opened up markets for artisanal goods and services. Carters took cane crops to mills fueled by wood provided by woodcutters. Weavers of bamboo baskets and mats as well as rope, brick, and tile-makers all saw growing markets for their goods under the Cultivation System. As milling became more capital-intensive, it moved into the hands of Chinese-Indonesian, and then European, owners, reinforcing socioeconomic disparities along ethnic lines.[24]

After 1870, the Cultivation System gave way to private sugar planting, with foreign planters leasing land for their sugar crops from local farmers, usually on a three-year rotation with rice. Plantations used wage labor and imported sophisticated techniques and technology for sugar production, including new cane varieties, fertilization and watering techniques, and up-to-date milling technology.[25] Planter associations and the colonial government both supported these endeavors. Yet just as under the Cultivation System, participation by Javanese farmers as lessors of their rice land and occasionally as laborers was often negotiated by village heads, making it less than fully voluntary. By the early twentieth century, anti-colonial activists regularly decried the inequity of the typical contracts.[26] Lease payments to Javanese farmers were based on rice profits foregone, rather than a percentage of the sugar profits, reinforcing sugar profitability even when global sugar prices declined.

The rotation between rice and sugar complicated agricultural practice. Water became a scarcer resource, especially as sugar-planting techniques increased the need for carefully managed water provision. The government instituted day-night regulations in response to water shortages, giving sugar producers water during the day and leaving food growers to work in unlit fields at night. Alternative technical arrangements like cisterns that stored nighttime water for day use improved the situation, although silt buildup added heavy maintenance responsibilities to villages. Sugar cultivation exhausted the soil, making subsequent rice crops increasingly vulnerable to disease. By contrast, flooded rice often brought nutrient-rich silt onto the fields, essentially subsidizing sugar production by the improved fertility of the soil provided by the rotation with rice.[27]

Making export sugar profitable on Java rested first on Indonesian agricultural skill and, even when privatized, continued to rely on both Javanese labor and the technological skills of those farmers who reconfigured vital food production to accommodate the land lost to sugar. The sometimes contentious juxtaposition of sugar and domestic food production underscored the fundamental inequities of the colonial relationship.

Vietnam's Rubber Industry

In contrast to the preceding stories, the growth of rubber cultivation in twentieth-century Vietnam made little use of preexisting skills, although it did require considerable labor. Unlike familiar rice or sugar crops, *Hevea brasiliensis*, the rubber variety that dominated international markets, was new. Growing rubber required considerable capital (especially during the wait for trees to mature), often provided by foreign joint-stock companies, expertise derived from foreign sources, and wage labor to tap the trees and otherwise maintain the plantations. Rubber came to Vietnam (and other parts of Southeast Asia, especially Malaysia and Indonesia) as demand for rubber to make automobile tires skyrocketed (figure 38).[28]

The French colonial government approved of rubber planting but did little to support its introduction. Instead, planters and planter associa-

Figure 38. Large rubber plantation in Vietnam.
By Peter van der Sluijs via Wikimedia Commons. CC BY-SA 3.0, https://creativecommons.org
/licenses/by-sa/3.0/.

tions, whose mainly European membership included only a few Vietnamese or Chinese members, became the major players in introducing and developing the techniques and technology vital to Vietnam's rubber industry. Planters used the associations to share technical information about soil and water tolerance, disease, tapping techniques, and labor management, all essential for making the plantation system work. They especially shared critical knowledge from their colonial counterparts in Malaya and the Congo.[29]

Colonial officials did attempt to supplement foreign plantations with a smallholder system. They aimed to support smallholding entrepreneurs who could supply extra rubber when prices were high. Yet they failed to acknowledge rubber's formidable barriers to entry. Monocropped plantations are deeply vulnerable to plant diseases and insect infestations. New plant stock was essential to maintain yields as older cultivars succumbed, and all trees required continuous fertilization to maintain soil vitality. Expensive biotechnologies like grafted or cloned trees were necessary, as was a constant stream of information about experiences on rubber plantations elsewhere. The French colonial government provided little support in any of these areas. Research into soils and disease was usually available only in French, which smallholders might not speak or read. The one publication available in Vietnamese spoke to landholders with more than 50 hectares (a large holding) and focused more on labor management than the technical intricacies of rubber production. Colonial education programs likewise offered little attention to technical concerns. A Vietnamese or Chinese farmer hoping to create a small, viable rubber planting would be more or less on their own.[30] Unlike the Malay world, where seminomadic people (drawing in part on their knowledge of other latex-bearing plants) successfully integrated rubber cultivation into their systems of shifting cultivation, in Vietnam, the preference for intensive monocropping made it difficult for smallholders to succeed.[31]

These cases highlight a few of the diverse ways agricultural expansion worked in nineteenth-century Southeast Asia, although cases like the growth of Vietnamese rubber seem iconic, with foreign technical knowledge, Western markets, and big capital leaving local people to participate mainly through their labor. Yet rice and sugar expansion in Burma, the Philippines, and Java shows the leading role local technical knowledge and experience played in developing export commodities. Local farming systems were no relics, doomed to be overwritten by incoming technologies and techniques, as colonial authorities sometimes claimed. By managing production costs and leveraging local technical knowledge, Southeast Asian agricultural practices were as valuable for expanding agriculture in the nineteenth century as they had been five centuries earlier. Colonial efforts to "improve" agriculture with new fertilizers or seeds often aimed to extend the viability of local practices, not transform them.

However, expanding agricultural production could generate new social, environmental, and technical assemblages. Farmers transformed Burma's south incrementally through land reclamation and flood protection techniques to support monocropped rice. The integration of sugar and rice production on Java introduced new technical challenges for rice growers, even as their practices ecologically subsidized sugar.

Incoming technologies and distant markets mattered, but they should not make us overlook the importance of Southeast Asian techniques and technology in nineteenth-century change.

Foraging Peoples and Natural Commodities

The exploitation of forest and sea products expanded significantly, especially as products like rubber and gutta-percha became vital to modern industrial products and infrastructures, and existing markets for natural commodities like birds' nests, sea cucumber, and rattan remained strong. European and Southeast Asian actors sought to monopolize trade in these products. The British made first Penang and then Singapore the main destination for Chinese and Indian country traders to sell the goods they obtained from forest collectors. The Tausug Sultanate of Sulu, exerting political control in much of North Borneo and the Sulu Archipelago from 1780 to 1860, controlled trade in marine and forest goods by granting princes rights to collect goods from villages.[32] As was true in earlier centuries, foraging communities and enslaved laborers did the actual work of collecting. Although techniques for extraction may have changed only modestly, the growing demand for these commodities had real consequences for the technological lives of foraging communities. For example, compelled labor among seagoing peoples in Sulu gradually eliminated their subsistence activities (fishing and coastal agriculture), changing their technological and economic ways of life.[33]

Colonial authorities also attempted to use the technological labor of foraging peoples. Europeans typically disapproved of swiddening, seeing it as "wasteful," and thus couched efforts to eliminate swiddening as a form of uplift. For example, British authorities hoped to transform Karen seminomadic swiddening peoples into foresters in colonial Burma.[34] In the early years of British rule, uncontrolled logging had devastated valuable teak forests. The British looked to the Karen practice of *taungya* forestry, a mode of exploiting and husbanding forest products (although not teak), as the answer. The British encouraged Karen cultivators to plant and care for teak saplings on some of the land cleared for rice and cotton production. Karen people initially refused, however, because the presence of teak would make it impossible for them to return and clear these lands later on, rendering their swiddening practices impossible—something the British understood, too. Officials applied pressure, including by mandating teak planting as a punishment for "misuse" of state-controlled forest land. Ultimately, after promising the Karen people designated spaces within forest reserves for shifting cultivation, the Karen acquiesced. They took up teak cultivation and provided forest services like fire-fighting, compromising their independence and modifying their sociotechnical ways of life.[35]

Unquestionably, foreign markets and foreign technologies drove intensive exploitation of forest products. European adventurers were often on the lookout for valuable natural products in their colonial holdings. Consider the widely studied case of gutta-percha. In the mid-nineteenth century, European companies attempting to install undersea telegraph cables discovered that the latex of several varieties of Southeast Asian trees, the *Sapotaceae*, commonly found and used in the Malay and Philippine Archipelagos, could act as an ideal insulator. On sending a sample to England from Singapore in 1843, Charles Nickles noted that it was waterproof, moldable when heated but hardened after cooling, and resistant to chemicals.[36] As demand for gutta-percha took off, forest collectors traded with villages and Chinese or Indian country traders, with much of the trade moving through Singapore. European expeditions lacked the intimate knowledge of the landscape to reliably find the widely spaced trees. Efforts to create plantations, such as Tjipetir in the Netherlands East Indies, failed.[37] The *Sapotaceae* yield poorly when tapped and are usually just cut down. From a financial standpoint, plantation models did not pay.

Instead, the unacknowledged technical knowledge and skills of forest collectors and the processing techniques employed by Chinese and Malay traders filled the growing demand for gutta-percha.[38] Forest collectors embarked on lengthy and dangerous expeditions to find widely spaced trees, not all of which were the same species. Seminomadic peoples like the Iban Dayak peoples of Borneo were ideal collectors.[39] Working in the agricultural off-season, collectors accrued both economic

Figure 39. Four Kayan men collect gutta-percha from a tree trunk in Sarawak.
Courtesy Wellcome Images. Public domain.

benefits and the social respect that such expeditions (which became longer as trees became rarer) engendered (figure 39).[40]

Because it was so profitable, gutta-percha collecting may at times have disrupted rice production (or removed too many hands from the work), as evidenced by increasing rice imports into gutta-percha collection areas, although swiddening was never abandoned.[41] European officials condemned the practice of cutting down trees, but with no other way to obtain the latex, they accomplished little more than reinforcing cultural narratives that linked swiddening and collecting to inefficiency, laziness, and shortsightedness—something for colonial civilizing missions to eradicate. Yet, as with agriculture, foreigners and Southeast Asians alike built their business models on the continuing existence of collectors and the techniques and knowledge they provided. Exports of gutta-percha from Singapore and Sarawak trended upward into the twentieth century, with more than 6,500 metric tons of gutta-percha exported from Singapore in 1901 alone.[42] Demand for gutta-percha did not decline until the mid-twentieth-century development of synthetic thermoplastics. Although colonial histories tend to dismiss collecting cultures as relics of premodern times, they remained essential to Southeast Asian economies. Their skills and knowledge played a critical role in making distant modern technologies viable.

Infrastructure

The nineteenth and early twentieth centuries saw aggressive efforts by governments and private entrepreneurs to transform major infrastructure as an essential part of reconfiguring Southeast Asia's political and economic order. Both established infrastructure such as ports, canals, and irrigation works, and new technologies like railroads and motorways featured significantly in "modernizing" projects intended to expand, intensify, or streamline the production and flow of Southeast Asia's increasingly valuable commodities. Those same projects also inspired Southeast Asians to weigh in on and sometimes contest the sociotechnical futures that enthusiasts aspired to materialize.

Infrastructure consists of the "basic facilities, services, and installations needed for the functioning of a community or society."[43] Yet infrastructures do not serve all equally, offering both affordances and barriers to desired activities.[44] Advocates of new or revitalized infrastructure create imaginaries of sociotechnical functioning, making them potent sites for political contestation. Both new and existing infrastructures in Southeast Asia were embedded in discourses about the character of modernity. But the meanings of technological modernity for Southeast Asia were worked out locally, in practice, not determined in advance. Although scholarly literature often focuses on the power of infrastructural technologies, especially "transferred" technologies, a closer look at both old and new infrastructure also highlights their risks, fragilities, and fluid meanings.

Transforming Waterscapes

Water infrastructures for transport and agriculture continued to be vital to Southeast Asian life. Transformation of water infrastructure

BANGKOK: "THE VENICE OF THE EAST."

Figure 40. Bangkok, c. 1907.
In Henry Norman, *The Peoples and Politics of the Far East.* Public domain.

often involved expansion, especially of ports, to accommodate new steamship technologies.[45] Although some decried the slowness of water transportation compared with new rail- or roadways, water still carried a large portion of the goods and people that moved around the region. For example, people in Bangkok used canals as their primary means of transport around the city until the early twentieth century. Bangkok's gradual introduction of paved roads was almost incidental rather than a self-conscious effort at modernization.[46] Until 1890, the increasing quantities of goods passing through the city traveled by water, motivating improvements to canals, not roads (figure 40).

Water transport made sense in the swampy delta region. Paved roads appeared mainly around royal residences, as feeders to canals, and later in European districts where residents asked for roads for health reasons. After 1890, a more focused road-building strategy emerged because of economic expansion and population growth driven by Chinese immigration. As the number of businesses grew, more roads fed the canals, and more residences housed a larger permanent workforce than the city had ever seen. After the introduction of automobiles, Siamese royalty saw roadways as a promising business opportunity, introducing improved maintenance and construction technologies (including steam rollers) in the process.[47] Traditional waterways and new roadways both served Bangkok's economic transformation.

When the French colonial government undertook water projects in the Mekong Delta Region, their aims were more explicitly political.[48] Flat geography, regular flooding, and shifting sandbars made the Mekong a tricky site for navigation. The delta had long served as a refuge for rebels against sitting governments; anti-colonial activity in the later nineteenth century was just the latest example. The French first built inland canals to move their steamships to the delta's swampy sites of rebellious activity. But the delta defeated French ambitions. Flooding, necessary for rice production, created sandbars in the new waterways. Later, the French employed dredging equipment to deepen channels and extend waterways. Chosen as a cheaper and more humane alternative to corvée labor, dredges were nevertheless expensive and subject to frequent breakdowns, requiring constant, costly maintenance.[49] Villagers sometimes sabotaged projects that obstructed desperately needed floodwaters. Yet waterways dug with steam dredges ultimately transformed the delta environment, devastating existing farms and bringing migrants (who used different farming practices) to clear the forests and establish settlements.[50] Intended to show the benevolence of colonial rule, they instead created an unpredictable landscape of disruption and opportunity. Farmers had no guarantee that their reclaimed land was ecologically or economically viable; the delta region suffered serious agricultural crises during the difficult 1930s, with some facing total bankruptcy.[51] Costly new water infrastructure failed to materialize the French vision of interconnection, productivity, and social harmony. Instead, it created a risky new envirotechnical landscape that provoked as much disaffection as it did collaboration.

Railways

Water infrastructure, however modernized, reinforced a long-established hydraulic logic in Southeast Asia. In contrast, railways exemplified a new, land-based sociotechnical imaginary, which Sheila Jasanoff and Sang-Hyun Kim define as a collectively shared vision of desirable futures defined by particular sociotechnical arrangements.[52] Globally, governments and businesses saw in railways an opportunity to assert political and commercial control and bring remote areas into expanding webs of commodity production. Southeast Asian railway planners shared the same dreams, hoping to increase opportunities by "shrinking" distance. Yet they refracted these broader imaginaries through the lens of Southeast Asian political, economic, and social circumstances.

Commercial aspirations for Southeast Asian railways often involved speeding up or streamlining commodity transport, in the process advocating new ideas about productivity and efficiency. Far from mere instruments to achieve preexisting goals, technical affordances, relationships with existing infrastructure, social value, and cost shaped rail projects. For example, on the tin-mining island of Bangka, rail lines provided an alternative to cartage when pollution from mine tailings left rivers unnavigable.[53] On Java and Sumatra, rail projects promised to quickly transport crops like sugar, coffee, and tobacco to ports, even during the dry season when water transport was difficult. It otherwise took five months for crops to be hauled over roads from central Java to Surabaya or Batavia.[54] Some rail advocates went further, portraying the railway as a way to make scarce labor more mobile and productive. One editorial writer in the 1840s saw railways as a key way to expand production of exportable commodities, portraying this as "productive labor" in contrast to the "unproductive drudgery"—like growing food—that people would otherwise do. Furthermore, he (wrongly) portrayed railways as easy to construct, an almost providentially natural fit for Java's geography. Thus the imaginary encompassed more than speed; it asserted a radical revision of the social meaning of productivity and a view of railways as an easy and natural "fit," nestling painlessly into the landscape. Although it was not until the late 1880s that Java had completed its eight railway lines, two generations of advocacy suggest that imaginaries like these had real appeal.

Other railway projects aspired to create, rather than recreate, productive landscapes. In Sabah (North Borneo), railroad promoters in the

British North Borneo Chartered Company suggested that a rail system would drive new production of valuable commodities in the region.[55] In the mid-1800s, the British used the island of Labuan as a coaling stop on their way to China, sparking wider interest in the region. British investors aimed to combine plantation development with trade in forest products. The east coast was already well served by a river system, long used by local and Chinese country traders to collect commodities. However, the west coast had only short rivers, with locals more focused on rice production than trade. The British viewed the interior areas of the west as unpopulated, although it is likely that nomadic and seminomadic people circulated through the region.[56]

Advocates hoped railways in the west would mimic the river access of the east, making cultivable land more easily accessible to planters and transporting Chinese immigrant laborers to serve as the workforce.[57] W. C. Cowie, a shareholder in the British North Borneo company, advocated tirelessly for a railway, convinced of its potential by Frank Swettenham's success in building a railroad to carry tin from Kuala Lumpur to the port of Klang in the Malay states. Cowie eventually persuaded the company to fund the project without, it appears, significant planning, surveying, or engineering study. Work started in 1896 but faced serious obstacles from geography by turns mountainous and swampy. The lack of advanced planning of viable sites and ports and significant engineering blunders dogged the project. Financially, too, Cowie's aspirations required a leap of faith. Extending rails into the interior made little sense; the lightly populated areas offered few paying customers.[58]

The North Borneo rail project would have failed completely and expensively were it not for the timely intervention of the rubber boom of the early twentieth century. Aspiring rubber planters received good terms from the company, who even cleared land on their behalf. Migrant railway laborers from Java and China received incentives, including land near the rail line, to build and maintain the rail system; some later opened sawmills or became farmers. By the end of the 1920s, foreign rubber planters and migrant laborers had fulfilled the visions behind this risky and poorly planned railway. Indeed, the growth of rubber planting made it possible to reengineer the most poorly designed parts of the system. Before the rubber market collapse in the late 1920s, the railroad project looked like a smart play (less so afterward). Yet the North Borneo railway project was neither well planned nor well executed, an accidental success that underscores the fragility of aspirational infrastructural projects, which need to configure many technological elements, in addition to rails themselves, to work. The railway was no intrinsically powerful technology masterfully wielded by Western colonizers. Only the contingency of the rubber boom saved it.

Railways appealed for the often intertwined reasons of military, political, and commercial influence. In Vietnam, the French colonial government constructed the Indochina-Yunnan railway in hopes of gaining a foothold in the China trade to rival what the British enjoyed in Hong Kong, a forlorn hope.[59] The Siamese monarchy likewise hoped its first rail line, between Bangkok and the arid Khorat Plateau, would stimulate new commodity production in the region and allow them to fend off aggression on their distant borders.[60] A state address by King Rama V hints at tightly interconnected commercial and political aspirations embedded in railway construction: "We are convinced that, to a very large and important degree, the material progress and prosperity of a people usually depends upon its means of transport. When there are good means of transport, people can travel easily and quickly over long distances. The population will be enlarged. Commerce, the foundation of a country's wealth, will prosper. We have therefore been diligently striving to build a railroad befitting the strength of Our country."[61]

On the mainland, railway construction was often taken up in the hopes of either spreading or resisting colonial expansion (formal or informal). In island Southeast Asia, political expansionism was less common, although most endorsed the value of rails for moving troops. Advocates of the Indochina-Yunnan line, for example, stressed its military potential to put down resistance in the North Tonkin. Others suggested that access to the port city of Kunming in Yunnan would make it possible to establish de facto French control over the whole province. Ultimately, France neither annexed Yunnan nor faced any serious opposition that the rail link to Kunming could have served. As a "tool of empire," the Indochina-Yunnan railway accomplished next to nothing.[62]

Having said that, it is important not to overlook the real affordances of a working railway line. Providing physical access to areas not already served (and possibly protected) by waterways could certainly play a role in changing the balance of power in a region. Nowhere was this more evident than in Siam, which faced aggressive moves from both British and French players eager to gain commercial and therefore political leverage in the last uncolonized area of Southeast Asia. In Siam, government officials viewed the construction and design of railway lines as existentially significant. Rail lines offered multiple entry points to Siam for those seeking control; the Siamese government had to be strategic about where and how to run rail lines. As the British and French plotted imperial expansion, the Siamese used railways to deflect the most aggressive moves of would-be colonizers. From the 1880s, British, French, and German entrepreneurs attempted to gain railway concessions to control trade, and through trade, Siam. For example, British efforts to build railways on the southern border between Siam and British Malaya were transparent attempts to flood the region with British goods, integrating the region into Malaya in ways that would limit Siamese control. Siamese leaders likewise rejected British suggestions for rail lines that integrated Siam with British Burma.[63] The Siamese had no desire to make entry across their borders easier.

Yet the Siamese also saw the strategic possibilities in railways. When the French angled to annex Lao territories claimed by the Siamese as vassals, the Siamese government considered running a rail line to more aggressively hold those areas; the plans were abandoned as the French turned their attention elsewhere. With more subtle challenges, they chose to appear accommodating. For example, they provided the desired British concession in Perak on the border with British Malaya but refused to provide income guarantees. The British concessionaires could not find investors willing to take on such a risky project.[64]

Ultimately, however, the Siamese government chose to create a state-run railway system, with some foreign financing permitted subject to the direct approval of the monarch. Although a state-run system foreclosed aggressive attempts to win railway concessions, this move did not solve all of their problems. Because of the scale and complexity of the technology, railways were expensive. A poorly chosen route could bankrupt the state. Conflicting internal priorities also shaped the rail system. Military officials advocated for a narrow-gauge railway because they thought (incorrectly) that it would differ from Burmese railways, making them more secure from invasion. But they also worried that the wide-gauge railroads that the ministry wanted would deplete the military budget.[65]

Controlling railway construction did not by itself promise full autonomy. The Siamese also faced dependence on foreign experts to operate and maintain the system. The Siamese monarchs admired the Japanese model of training local experts to reduce dependency on foreign experts to run the system, but they never produced a similar result at home. Although Siam's royalty and aristocrats were encouraged to attend foreign universities, technical education was usually seen as beneath them. One exception to this was Prince Phurachatra, who studied engineering in England before returning home to command the Army Engineers.[66] Others, for whom technical education would have been appropriate, did not receive funding for expensive foreign travel. In lower schools, debates between traditionalists and modernizers slowed efforts to introduce more science and technology education.[67] Therefore, despite Siam's defensive view of railways, it was easier to control the material infrastructure than create a deep foundation of expertise that would have given them more complete autonomy.

Imagining Modernity

As elsewhere, infrastructure projects in Southeast Asia became sites where ideas about modernity and modern identities were formed. Modernity is therefore not a singular or entirely shared experience. Rather a multiplicity of modernities emerged as a result of divergent experiences. Infrastructure projects inspired arguably new forms of social action in Southeast Asia and introspection about the deeper meaning of these changes. However, such responses were embedded not just in the technology alone but in the wider political contexts and in diverse visions of the future.

A prominent enthusiast of modern technology was the Javanese aristocrat Raden Ajeng Kartini. An expressive and prolific correspondent with her international circle of friends, she commented in optimistic terms about the future of the Netherlands Indies, sprinkling her writing with references to the freedom that technologies like airplanes and railroads could offer.[68] She saw such technologies as a way of giving the Javanese people access to the rest of the world. For all of her optimism, however, Kartini's work also speaks to the dissatisfactions they could produce. Her longing for an airplane ride was never fulfilled. She often wrote of train and tram rides, which she enjoyed during the rare times she was permitted to travel. But the possibilities they offered for connecting with distant friends (even if the train also speeded the end of such visits) were perhaps the most appealing: "Do not fly so fast on the smooth iron tracks, you sniffling, steaming monster, do not let this beautiful meeting end so quickly. . . . I prayed that the ride would never end."[69] The intense joy she experienced in riding the train emphasized the routine lack of mobility that characterized her ordinary life. For Kartini, modern technology represented not just pragmatism but the possibility of new kind of personal liberation she hoped all Javanese people could share. While government officials in charge of building railways described them mainly in terms of defense or trade, passengers like Kartini found the possibilities of individual mobility just as compelling and just as indispensable, once experienced (figure 41).

Elsewhere, too, as in much of the rest of the world, passengers flocked to rails. In Siam, for example, a line built from Bangkok intended to improve the shipment of vegetables into the capital was instead flooded with passengers.[70] The ability to travel more frequently and in many cases faster even than by water offered novel experiences that could shape wider perspectives. Rama V saw the Khorat Plateau of Northern Siam for the first time in his life on a newly established train route to that arid region.[71]

Workers on railways also gained the chance for new experiences on the railway. Doing everything from line maintenance to serving on train cars or in stations, they often enjoyed relatively well-paid jobs and gained opportunities to travel.[72] In Vietnam, for example, although rail workers

Figure 41. A train crossing a bridge near Malang, on Java. Images like this one highlighted the ways the railroad could connect people across natural barriers. Courtesy Leiden University Digital Collections. Shelfmark KITLV 15464. CC BY-SA 4.0, https://creativecommons.org/licenses/by-sa/4.0/.

frequently complained about management, hours, and working conditions, they spoke positively about the access to technical education that working on the railway offered them.[73] Railway workers gained access to colonial schools for their children and colonial health care, as well as the opportunity to travel. The railways did not suddenly make these things desirable, but they did put them within reach of more people. Those people very likely carried these new perspectives with them to their home villages, a result satisfying to French authorities who sought to "modernize" Vietnamese peoples.[74]

Yet railways also became a new site for colonized peoples to experience inequalities along lines of gender, ethnicity, class, and colonial privilege. Although nothing new, the contrast between the promises of modernity embedded in colonial rhetoric and the reality that even on the modern new railways old social hardships prevailed made a powerful case against colonialism. On Java, Malay passengers could be forced to wait outdoors to meet friends or family while Europeans or Chinese would be accommodated in waiting rooms.[75] In Vietnam, Governor-General Paul Doumer built the Transindochinois railway as a platform for associationism: the idea that Vietnamese, Cambodian, Lao, European, and Chinese people would find the opportunity to work together for collective economic betterment. Yet the railways highlighted the anti-associationist tensions implicit in colonial life as time went by. As Southeast Asian ridership increased, Europeans took to the roads in private automobiles instead, seeking segregation in their choice of mobility.[76] Railway workers engaged in modern actions like strikes; whether intended as an anti-colonial protest as on Java[77] or as a response to working conditions or wages as in Vietnam,[78] they reflected a growing understanding of the power of collective action at such sites. Disrupting the smooth running of infrastructure called into question both the pragmatic power of colonial authority and the validity of the sociotechnical imaginaries that justified them. These technologies might be used to spread colonial control in certain ways, but they could equally be used to undermine it.

Conclusion: Technological Transformations

In Southeast Asia, as in other parts of the colonized world, Europeans developed polarizing rhetoric about technology, asserting the technologies as most modern and meaningful as foreign "gifts" to the peoples of Southeast Asia. Even in uncolonized Thailand, civilizing rhetorics embraced an ideal of the (implicitly superior) new as foreign, and the old as local.[79] Figures like Kartini evinced a longing for the novel modes of connection that incoming technologies offered, connections that gave new opportunities for good lives. Other incoming technologies, which limited space precludes us from exploring, like radios, photography, film, and automobiles, also evoked enthusiastic responses, the broader consequences of which remain to be explored for much of Southeast Asia. By the early twentieth century, dissatisfaction with the official view of colonial economies as sites for extraction rather than manufacture and limited access to technological training emerged as a recurrent theme in anti-colonial thinking.[80] Even here, the focus on technology as something foreign and withheld prevailed.

Yet, in reality, Southeast Asian technological practices retained a vital, ongoing role in Southeast Asian economies, including their crucial role in propping up incoming technologies. European traders and colonizers confronted a technological landscape in Southeast Asia that had long been characterized by opportunistic technological exchanges in contexts of both equal and unequal power. They encountered and depended on skilled technological agents who might work cooperatively or oppose European desires (or both). These realities have often been obscured by scholarly and political attention to potentially transformative incoming technologies, resulting in a tendency to depict Southeast Asia as a place of absence, technologically speaking. More research on how Southeast Asians defined or used the term "technology"—which was only coming into international use in the early twentieth century—could shed light on what aspects of this experience seemed to Southeast Asians to be a break from the past, and which past. After all, perpetual circulation and innovation of technologies and techniques were normal in many parts of Southeast Asia for millennia. We know something about how Southeast Asians felt the modern world to be different from the past. Is there more to be said about the ways it may have felt the same?

For all of the attention given to how technologies strengthened the imperial hand, the history of technology in nineteenth-century Southeast Asia reveals as much about imperial weakness as it does about strength. The seemingly counterintuitive idea of technological weakness is a vital dimension of colonial history. Without dismissing the important consequences of colonizing interventions, Europe's ascendancy in Southeast Asia was relatively brief. Rather than seeing European technological strength as an explanation of colonial strength, or a tool of empire, it is more useful to see technologies as sites of ongoing and indeterminate processes of reconfiguration, negotiation, and struggle over the region's future. As analysts, this perspective encourages us to see all the technology in this picture, both incoming and preexisting, and to seek a better understanding of technological transformation in the assemblages rather than the individual technologies. Doing this, we may find the power of technology to be far more slippery and evanescent than it seems.

Conclusion

EXPLORING SOUTHEAST ASIAN HISTORY over the longue durée underscores how technologies were co-constituted with diverse yet interconnected cultures, facilitating and motivating patterns of circulation central to its life as a region. Going beyond descriptions of the merely instrumental uses of technologies, I have highlighted, wherever possible, the profound entanglements of technology with politics, social relationships, and culture. People used technologies and techniques such as ships, looms, rice terraces, and swiddening to achieve specific practical goals. But those same technologies could establish social solidarity, define or challenge social hierarchies, or inform cooperative or conflictual relationships with distant peoples.

Throughout Southeast Asia's history, both indigenous technological innovation and incoming technologies matter for understanding the region's characteristic form of technological dynamism. Southeast Asian societies both created novel technologies and took in foreign technologies, mixing and reassembling them in ways both disruptive and accommodating to preexisting ways of life. The technological politics of Southeast Asia since the mid-twentieth century, a politics that has emphasized economic development, technology transfer, and global integration, can lead scholars to portray Southeast Asia as a negative space, technologically speaking, shaped primarily by incoming technologies. Yet viewed in a longer historical perspective, the recent history of Southeast Asia's technological transformations, although unique to their time, is nevertheless just the latest iteration of a process that Southeast Asians have engaged in and managed for centuries. Invention and localization, assemblage and hybridization are all interconnected. Each of these different modes of technological action has implications for the sociotechnical order that Southeast Asians have produced and reproduced. The rich possibilities of the history of technology as an analytic emerge when we focus more on the evolving practices of technological engagement and the wider sociotechnical processes that govern technological stability and change.

Two themes in this synthesis resonate from the deep past to the present day in Southeast Asia, the significance of technological intensification and expansion of production for the exercise of power, and the role of technology in the formation of social, commercial, and political interdependencies within and beyond the region. Both of these themes have received critical attention from historians working on the colonial period, but they are also clearly evident from ancient times up to the present day. Expanding production and extraction, whether by adopting new techniques and labor strategies or through voluntary or involuntary migrations, has been central to shifting power relations in both states and in communities. The consequences of these choices have been political, social, and ecological, shaping contemporary landscapes, business practices, and state action. Efforts to expand production today, although different in the particulars, nevertheless resembled approaches pursued over the past five hundred years. Disputes over land use and the wisdom of agricultural or industrial transformations of uncultivated land inspire debate based both on modern concerns like climate change and long-term, entrenched tensions between settled and nomadic ways of life. In all of these stories, technologies themselves, their affordances and requirements, are inescapably entangled with social and political conflict.

Likewise, technology was deeply implicated in anxieties about interdependency between states. Skilled technical experts, especially in the years when Southeast Asia was thinly populated, were a boon to any state. Their abilities enhanced a state's autonomy and resilience to change. Contemporary concerns about forms of technological dependence play a role in both international diplomacy and economic decision making in ways that resonate with the concerns of the past. Modern economic policies, although informed by modern ideas about "economic development," are fundamentally concerned with the balance of local and imported technologies and skills. Their anxieties are directly informed by the experiences of recent history, and indirectly by deeper historical relationships and patterns of circulation.

Southeast Asianists have always made a place in their work for stories about technology, even when technology per se was not their central focus. Studying Southeast Asia's technology history in ways that connect the premodern to the modern is to embrace interdisciplinarity and the myriad insights it provides. Technology history can build on and engage with these stories by opening them up to new questions, exploring long-term sociotechnical continuities as well as moments of disruptive transformation, and taking seriously the interconnected practices of making, knowing, and doing that have characterized Southeast Asian history.

One of the great pleasures of finishing a book is having the opportunity to thank all who helped bring it to fruition. My biggest thanks go to Pam Long and Asif Siddiqi, the series editors for Johns Hopkins University Press, whose generosity with their time and expertise was instrumental to the completion of this project. I am extraordinarily grateful for their unflagging advocacy for the topic and sharp critical readings of drafts, which made this book stronger than it would otherwise have been.

I am also indebted to the anonymous readers. Their constructive criticism supported the work even as it pushed me to think harder about bringing technology studies and Southeast Asian studies together in a way that speaks to readers coming from different starting points. Likewise, I thank my colleagues Hunter Heyck, Peter Soppelsa, and Stephen Weldon in the Department of History of Science, Technology, and Medicine at the University of Oklahoma. From administrative support, to patient readings of the odd chapter draft, to informal hallway conversations that veered off into Southeast Asian waters, they offered input and perspectives I deeply value. Librarians JoAnn Palmeri and Melissa Rickman helped me find and scan beautiful images from books in OU's remarkable History of Science collections. I thank the University of Oklahoma for travel and writing support at various times during the past several years. I also thank the undergraduates in my Science and Technology in Asian History class, who read early versions of several chapters and, through their reflections and assignments, helped me see this work in new ways. You didn't know you were helping, but you did.

I am grateful to my many colleagues working in the dynamic technology field in Asian history and culture. Through formal and informal conversations at numerous conferences and workshops, many of you have lent a willing ear and shared your insights and points of view. I am particularly indebted to Francesca Bray, Aleksandra Kobiljski, and Dagmar Schaefer. Their explorations of technology in Asian history and tireless efforts to build a vibrant intellectual community around this topic have been a continuing source of inspiration.

I am more indebted than I can say to the many scholars who have dedicated their careers to scholarship on the histories and cultures of Southeast Asia. Digging deeply into this diverse and sophisticated scholarship was a constant joy—

and I still feel I only scratched the surface. I hope this book, in some small way, will inspire young scholars to embrace this wonderful field of study and perhaps add more to the story of technology in Southeast Asian history.

And finally, a special thanks to family and friends (you probably aren't reading this, but if you do, you know who you are) who put up with periodic, inexplicable conversational sidetracks into obscure bits of Southeast Asian technology history. Your support grounded me. Thank you!

amok. A raging or berserk form of attack, from the Malay.

Angkor. The capital of the Khmer Empire (802–1351 CE), also sometimes referred to as the Kingdom of Angkor, and home to Angkor Wat, the central temple of the city. The city was also known as Yasodharapura.

Arakan. A coastal region of present-day Myanmar, it was home to the Kingdom of Mrauk-U (1429–1785 CE). It is now referred to as Rakhine.

Austroasiatic languages. A primary family of languages found in mainland Southeast Asia as well as parts of China, India, Nepal, and Bangladesh. They are also referred to as Mon-Khmer languages.

Austronesian languages. A primary family of languages found in island Southeast Asia and parts of the mainland, the Pacific Islands, and Taiwan.

Ayeyarwady. Also transliterated as Irrawaddy, the largest river in Myanmar, running north-south and emptying into the Andaman Sea.

Bagan. Also transliterated as Pagan, this refers to the city and the kingdom. The kingdom was established in the ninth century and fell in 1287. The city has remained up to the present day.

Bago. Also transliterated as Pegu, it was a major port city in Myanmar. The date of its founding is controversial, but it was established no later than the late thirteenth century.

Bajau. Also Sama-Bajau, a group of Austronesian-speaking maritime peoples who originated in the Sulu Archipelago in the Philippines. The term is also transliterated as Bajo, Badjao, and Bayao.

Banaue rice terraces. Rice terraces in Ifugao, Philippines, dating from probably the middle of the thirteenth century.

Banda Islands. A group of islands in the region of Maluku in Eastern Indonesia. These were the only islands from which the spices of nutmeg and mace could be obtained.

Banjarmasin. A city on the island of Borneo, in the Indonesian province of Kalimantan, founded in 1526.

Borobudur. A Buddhist temple located in central Java, constructed in the ninth century.

brigantine. A sailing ship with two masts, with sails rigged to be perpendicular to the ship's hull. The Dutch word for this ship is *brigantijn*.

Champa. A mandala-style kingdom in what is now southeastern Vietnam, it lasted from approximately the second century CE to 1832.

Chao Phraya. The largest river in Thailand and the most important for commerce. It runs north-south through the country.

chialoup. A type of sloop built and used in the Malay region in the early modern period. Design variants included both European and Malay characteristics. In English, it is referred to as a shallop. The word "chialoup" is probably from the French.

Đại Việt. A kingdom of Vietnam, from 1054 to 1400, and then from 1428 (after throwing off Chinese invaders) to 1804.

Dayak. A term referring to a culturally related but distinct set of Austronesian-speaking ethnic groups on the island of Borneo.

Đông Sơn. A name given to a Bronze Age people living in the Red River Valley area of Vietnam.

Funan. The Chinese name for a mandala kingdom whose name has not survived, located in what is now southern Cambodia and Vietnam. Dates are uncertain, but it was probably founded in the first century CE and ceased to exist in the sixth or seventh century.

Gia Long. The imperial name of Nguyễn Ánh, who became the first emperor of the Nguyễn dynasty in Vietnam in 1802 and reigned until 1820.

gonting. A type of ship used in the Malay world of uncertain design.

Hikayat Banjar. The chronicle of the kings of Banjarmasin, located in Indonesia on the island of Borneo.

Hmong-Mien. A family of languages (also called Miao-Yao) spoken in the northern areas of mainland Southeast Asia, especially in northern Vietnam and Laos.

Hoabhinian. Refers to a characteristic set of stone tools that date to the Paleolithic in Southeast Asia, from 10,000 to 2000 BCE. The term stemmed from the village of Hòa Bình in which the first finds of this type were unearthed.

hongi. An Ambonese term referring to a fleet of warships.

Iban. Refers to a Dayak ethnic group who live on the island of Borneo in Malaysia, Brunei, and Indonesia.

Ifugao. Refers to both a province in the Philippines in a mountainous region on the island of Luzon and to an Austronesian-speaking ethnic group who occupies that region.

Iranun. Refers to an ethnic group from Mindanao in the Philippines and Malaysia.

jong. An early Javanese ship design, also referred to as a junk, but with some differences in design from the Chinese ships also referred to as junks.

Kantu'. An ethnic group in Borneo.

karakoa. A Philippine sailing vessel used in warfare. It features outriggers and a deck to permit fighting at sea. Similar ships in other parts of maritime Southeast Asia have similar names, including kora-kora. Historical sources use numerous alternate spellings.

Karen. A Sino-Tibetan-speaking ethnolinguistic group made up of many independent, culturally distinct ethnicities located in southern Myanmar.

Khmer. A large Austroasiatic-speaking ethnic group in Cambodia.

Khmer Empire (802–1431 CE). An empire that covered the region that is now Cambodia and parts of southern Vietnam and Laos. The capital city was Angkor.

Khok Phanom Di. An important archaeological site in Thailand that was occupied between roughly 2000 and 1500 BCE.

Konbaung dynasty. A Burmese dynasty that ruled the country from 1752 to 1885.

kris. Also sometimes spelled "keris," a style of dagger found in the Malay world with a characteristic wavy blade.

Lan Na. A kingdom located in what is now northern Thailand, from 1292 to 1775.

Lê Dynasty. The dynasty that ruled Đại Việt between 1428 and 1789.

Majapahit. A Javanese mandala kingdom whose influence reached much of what is currently Malaysia and Indonesia from 1293 to 1517.

Makassar. A region of South Sulawesi now in Indonesia. Historically home to two major kingdoms, Talloq and Gowa, from the fifteenth century to the nineteenth century.

Malacca. Contemporary spelling is Melaka. A city on the straits of Melaka, which was a major trading power from 1396 to 1511, while ruled by the Sultanate of Malacca.

Maluku. Formerly referred to as the Moluccas. These islands are found in eastern Indonesia, between Sulawesi and Papua. Historically, they were called the "Spice Islands" by Europeans.

mandala state or kingship. A way of describing early Southeast Asian political organizations, where a central power exercised decreasing levels of authority over more distant areas, with some areas possibly included in the mandala of more than one ruler. Can also refer to polities organized as part of a federation.

Mataram Kingdom (716–1016 CE). Also known as the Medang Kingdom, it was an Indic-influenced kingdom that ruled large portions of central and eastern Java.

Mataram Sultanate (1587–1755 CE). An Islamic kingdom in central Java.

mayang. A Javanese boat with fore-and-aft-rigged sails and a characteristic upward turn in both stern and bow.

Mekong. A major river rising in the Tibetan Plateau and flowing through Myanmar, Laos, Thailand, and Vietnam.

Miao-Yao. *See* Hmong-Mien.

Mon-Khmer. A subfamily of the Austroasiatic, Mon-Khmer languages spoken in large sections of mainland Southeast Asia.

Mount Meru. The mythical center of the universe in Hindu and Buddhist cosmology.

Nguyễn Ánh. A Vietnamese aristocrat who defeated the rebel Tây Sơn dynasty and established the Nguyen dynasty.

Nguyễn dynasty. The dynasty started by Nguyễn Ánh in 1802 and lasting until 1945.

Niah Caves. Caves on the island of Borneo in the Malaysian state of Sarawak that house an important archaeological site with human remains from c. 40,000 years ago.

Nong Nor. An archaeological site in southern Thailand, occupied by hunter-gatherers, c. 2500 BCE.

Nusa Tenggara (Lesser Sunda Islands). A region of approximately 500 islands in the eastern portion of the Malay Archipelago, now in Indonesia.

Orang Asli. A Malay term meaning "original people," it is a catch-all category for numerous distinct non-Malay ethnic groups in Malaysia occupying the Malay Peninsula.

paduwakang. A type of traditional Malay sailing vessel that has two masts and a canted rectangular sail.

pencalang. A type of Malay ship originating in the Malay Peninsula region, with two masts and a continuous deck.

peranakan. A hybrid Chinese-Malay culture that first emerged in the Malay world in the fifteenth century.

perkenier. A Dutch system of silviculture, practiced in the Banda Islands.

Philippine Cordillera. A mountain range in central Luzon.

Punan. A catch-all term that is commonly used to refer to a large number of distinct and possibly unrelated ethnic groups in Borneo. There is also a distinct ethnic group called the Punan, who are swidden agriculturists on Borneo.

Pyu. People who established city-states starting around 200 BCE in what is now Myanmar. These states came under pressure from invaders from the north around 800 CE. The Pyu peoples were gradually absorbed into the Burmese kingdoms.

Rama I (r. 1782–1809 CE). The first king of the Chakri dynasty in Siam.

Semang. A nomadic ethnic group in peninsular Malaysia.

Shan. A Tai-speaking ethnic group in Myanmar, Thailand, and Laos. Many live in the Shan State in Myanmar.

Siam. The former name of the current nation of Thailand, founded in 1782.

Sino-Tibetan. A major language family that includes Burmese.

Srivijaya. A Buddhist mandala kingdom based on the island of Sumatra. Its capital was in the city of Palembang, 650–1377 CE.

Sulu Archipelago. Comprises islands in the Sulu Sea, in the southwestern portion of the Philippines.

swidden. A type of agriculture featuring short-term exploitation of plots of land, followed by long fallow periods.

Tai-Kadai. A language family found in mainland Southeast Asia. It includes the modern Thai language and many others.

Taksin (r. 1767–1782 CE). The King of Thonburi, the predecessor state to Siam.

Taungoo dynasty (1510–1752 CE). Also transliterated as Toungoo. A Burmese dynasty that ruled much of the region that makes up the current state of Myanmar and Thailand.

Tày. A Tai language spoken in northern Vietnam.

Tây Sơn. The aristocratic family who rebelled against both the Nguyễn and the Trinh lords, who were at war with each other.

tikus. Malay word meaning "mouse." It refers to a broker, often of Sino-Malay heritage, who would connect Malay rulers with Chinese mining entrepreneurs to exploit the rich ores of the Malay world.

Tonlé Sap. A large freshwater lake in Cambodia and the attached river. The lake is subject to seasonal inundation, making it an excellent location for rice agriculture.

Trần Hưng Đạo (1228–1300 CE). A prince of Đại Việt who commanded the kingdom's military forces. Among other achievements, he reformed military training.

Trịnh Lords. A family who ruled northern Vietnam between 1545 and 1787 under the authority of the Later Lê emperors.

VOC (Vereenigde Oost-Indische Compagnie). The United East India Company, usually known in English as the Dutch East India Company, was a chartered conglomeration of Dutch trading companies that operated in Southeast Asia as a trading company, but also had princely powers to engage in warfare.

xiangtou. A Chinese term denoting a mining expert responsible for managing engineering on mine sites such as the establishment of pumping apparatus to evacuate water from a mine.

Introduction

1. Victor Lieberman's *Strange Parallels 1: Integration on the Mainland* includes important discussions of the technological dimensions of economic change. Likewise, valuable descriptions of technology appear in Anthony Reid's *Southeast Asia in the Age of Commerce, 1450–1680.* Archaeological scholarship offers the most detailed analysis of Southeast Asia's technological culture in premodern times.

2. For the question of Southeast Asia as a region, see especially Lieberman, *Strange Parallels 1*; Reid, *Southeast Asia in the Age of Commerce*; and Sutherland, "Southeast Asian History and the Mediterranean Analogy."

3. Reid, *Southeast Asia in the Age of Commerce*, 1–11; Sutherland, "Southeast Asian History and the Mediterranean Analogy," 19.

4. I draw on Li Tana's insightful discussion of the analytic problem of indigeneity in Southeast Asian studies in Li and Cooke, *Water Frontier*, 5–8.

5. US Central Intelligence Agency, *The World Factbook.*

6. Bellwood, "From Prehistory to c. 1500," in *Cambridge History of Southeast Asia*, 1:55–136.

7. Association of Southeast Asian Nations, "ASEAN Statistical Yearbook, 2020."

8. Association of Southeast Asian Nations, "ASEAN Statistical Yearbook, 2020," 43.

9. Association of Southeast Asian Nations, "ASEAN Statistical Yearbook, 2020," 43.

10. Enfield, "Linguistic Diversity in Mainland Southeast Asia," in *Dynamics of Human Diversity*, 63–79.

11. The name Myanmar replaced Burma in 1989. Myanmar therefore refers to the modern nation-state.

12. In contemporary Myanmar, economic pressures may be threatening the viability of shifting cultivation in upland regions. Jepsen, Palm, and Bruun, "What Awaits Myanmar's Uplands Farmers?," 29.

13. "Myanmar," in McCoy, *Geo-Data*, 374–377. All entries from *Geo-Data* were accessed via Gale eBooks. Lieberman, *Strange Parallels 1*; Adas, *Burma Delta.*

14. "Thailand," in McCoy, *Geo-Data*, 533–536.

15. For a historical overview of language families in Southeast Asia, see Bellwood, "From Prehistory to c. 1500," 107–115. For more detail, see Peiros, *Comparative Linguistics in Southeast Asia.*

16. Georg, "Sino-Tibetan Languages," in *Encyclopedia of Modern Asia*, 234–236.

17. Diller, "Tai-Kadai Languages," in *Encyclopedia of Modern Asia*, 370–372.

18. Sidwell and Jacq, "Mon-Khmer Languages," in *Encyclopedia of Modern Asia*, 191–193.
19. Giersch, *Asian Borderlands*.
20. All data about contemporary religious affiliations is drawn from the Pew-Templeton Foundation, "Future of World Religions." Contemporary statistics provided by national governments may undercount minority religious groups. All information here is the best available but should be treated cautiously.
21. "Laos," in McCoy, *Geo-Data*, 304–307.
22. "Cambodia," in McCoy, *Geo-Data*, 89–92.
23. Clark, "Austronesian Languages," in *Encyclopedia of Modern Asia*, 190–192.
24. Hmong-Mien and Miao-Yao are both used to identify these languages. Georg, "Hmong-Mien Languages," in *Encyclopedia of Modern Asia*, 535–536.
25. "Vietnam," in McCoy, *Geo-Data*, 594–597.
26. "Indonesia," in McCoy, *Geo-Data*, 248–254; "Malaysia," in McCoy, *Geo-Data*, 333–337; "Singapore," in McCoy, *Geo-Data*, 482–484; "Brunei," in McCoy, *Geo-Data*, 80–81; Purdy, "East Timor," in *Encyclopedia of Environment and Society*, 374–377.
27. Foley, *The Papuan Languages of New Guinea*.
28. "The Philippines," in McCoy, *Geo-Data*, 432–436.

Chapter 1. Technology in the Human Settlement of Southeast Asia

1. See, for example, Lombard, *Le Carrefour Javanais*. Trade and circulation are treated extensively in Reid, *Southeast Asia in the Age of Commerce*, and Lieberman, *Strange Parallels 1*.
2. I have adopted Peter Bellwood's felicitous description of the results of human migrations into the region. Bellwood, *First Islanders*, 220.
3. The history of earlier hominins is beyond the scope of this book.
4. The oldest anatomically modern human skull so far found in Southeast Asia was uncovered in 2009–2010 in the Tam Pa Ling cave in Laos. Higham, "Hunter-Gatherers in Southeast Asia," 21–44. The date of 47,000–50,000 years is a conservative estimate.
5. Bellwood, *First Islanders*, 89–90.
6. For a detailed look at the diversity of evidence brought to bear on the biological history of this first migration, see Bellwood, *First Islanders*, 86–130, including Matsumura, Oxenham, Simanjuntak, and Yamagata, "Biological History of Southeast Asian Populations," 98–102. On the two-layer model, see Matsumura et al., "Craniometrics Reveal 'Two Layers' of Prehistoric Human Dispersal in Eastern Eurasia," 1–12.
7. On the term "Australo-Papuan," see Bellwood, *First Islanders*, 86.
8. Higham, "Hunter-Gatherers in Southeast Asia," 24–26.
9. Bellwood, *First Islanders*, 147–149.
10. Glover and Bellwood, *Southeast Asia*, 13–16.
11. Bellwood, *First Islanders*, 154.
12. Glover and Bellwood, *Southeast Asia*, 13–16.
13. Higham, "Hunter-Gatherers in Southeast Asia," 33.
14. Piper, "Changing Patterns in Hunting across Island Southeast Asia before the Neolithic," in Bellwood, *First Islanders*, 166–168.

15. Higham, "Hunter-Gatherers in Southeast Asia," 33–36.

16. Bellwood, *First Islanders*, 134–139; Ha Van Tan, "Hoabinhian and Before," 35–41; Higham, "Hunter-Gatherers in Southeast Asia," 25–31.

17. Bellwood, *First Islanders*, 134.

18. See Bellwood, *First Islanders*, for an overview of recent findings on technological assemblages and suggested readings.

19. Hung, "Neolithic Cultures in Southeast China, Taiwan, and Luzon," in Bellwood, *First Islanders*, 232–240.

20. Higham, "Mainland Southeast Asia from the Neolithic to the Iron Age," in Glover and Bellwood, *Southeast Asia*, 41–67.

21. Bellwood, *First Islanders*, 231–257.

22. Bellwood, *First Islanders*, 236. On the migration of farming communities, see Bellwood, *First Farmers*. For an alternative perspective on migration and settlement in Southeast Asia, see Solheim, Bulbeck, and Flavel, *Archaeology and Culture in Southeast Asia*.

23. Bellwood, *First Islanders*, 204–207.

24. See Castillo, "Rice in Thailand," 114–120.

25. Higham, "Mainland Southeast Asia from the Neolithic to the Iron Age."

26. For plants as technologies, see Moon, *Technology and Ethical Idealism*.

27. Castillo, "Rice in Thailand," 115–116.

28. Food and Agriculture Organization, FAOSTAT, http://www.fao.org/faostat/en/#data/QC. Millet, by comparison, is harvested on roughly 246,000 hectares.

29. Bellwood, "Origins and Dispersals of Agricultural Communities in Southeast Asia," in Glover and Bellwood, *Southeast Asia*, 21–40.

30. Higham, "Hunter-Gatherers in Southeast Asian History," 37–40.

31. Bellwood, *First Islanders*, 270.

32. Higham, "Hunter-Gatherers in Southeast Asian History," 40.

33. Bellwood, "Origins and Dispersals of Agricultural Communities," 34–35.

34. Higham, "Mainland Southeast Asia from the Neolithic to the Iron Age," 42–46.

35. Bellwood, *First Islanders*, 218–219, 268.

36. Higham, "Hunter-Gatherers in Mainland Southeast Asia," 39–40.

37. Mavhunga, *Transient Workspaces*; Soppelsa, "Intersections," 673–677.

38. Lemke, "Hunter-Gatherers and Archaeology," 3–18.

39. Lemke, "Hunter-Gatherers and Archaeology," esp. 7.

40. See also Mavhvunga's insightful analysis of technologies of the hunt in Zimbabwe in *Transient Workspaces*, esp. 41–71.

41. Kuchikura, "Efficiency and Focus of Blowpipe Hunting among Semaq Beri Hunter-Gatherers of Peninsular Malaysia," 271–305.

42. Baer, "Importance of Tools in Orang Asli Prehistory," 153–172.

43. Siebert, *Nature and Culture of Rattan*, 21–30.

44. Siebert, *Nature and Culture of Rattan*, 21–30.

45. Higham, "Mainland Southeast Asia from the Neolithic to the Iron Age," 51–64.

46. For a useful overview of swiddening as practiced around the world, see Mendoza, "Swidden."

47. See, for example, the consequences of pepper cultivation in Dove, "'Banana Tree at the Gate,'" 347–361.
48. Bellwood, *First Islanders*, 289–290. See also Castillo, "Rice in Thailand," for more discussion on the antiquity of dry rice cultivation.
49. Sellato, *Nomads of the Borneo Rainforest*, 11–13. For the politics of space among the Punan, see Großmann, "'Dayak, Wake Up,'" 1–28.
50. Shepherd and Palmer, "Modern Origins of Traditional Agriculture," 281–311.
51. Sellato, *Nomads of the Borneo Rainforest*. See also Moon, *Technology and Ethical Idealism*, 92–108.
52. Reyes, "Glimpsing Southeast Asian *Naturalia* in Global Trade, c. 300BCE–1600 CE," 96–119. Bellwood suggests that regular trade around island Southeast Asia started around 2000 BCE. "Southeast Asia Before History," 135.
53. Junker, *Raiding, Trading, and Feasting*.
54. Dallos, "Beyond Economic Gain," 403–422.

Chapter 2. Agriculture and Trade

1. See the useful comparison of different methodologies in Müller and Schurr, "Assemblage Thinking and Actor-Network Theory," 217–229.
2. Murphy and Stark, "Introduction," 333–340.
3. The South Asian subcontinent includes the present-day nations of India, Pakistan, Bangladesh, and Sri Lanka.
4. Stargardt, "From the Iron Age to Early Cities at Sri Ksetra and Beikthano, Myanmar," 344.
5. Bellwood, *First Islanders*, 223–226.
6. Bray, *Rice Economies*.
7. O'Connor, "Agricultural Change and Ethnic Succession," 968–996.
8. O'Connor, "Agricultural Change and Ethnic Succession," 972.
9. O'Connor, "Agricultural Change and Ethnic Succession."
10. Stargardt, "From Iron Age to Early Cities"; Lieberman, *Strange Parallels*, 89–90, 100–101.
11. Lieberman, *Strange Parallels*, 972. For more on farming practices in this region, see Higham, "At the Dawn of History," 418–437.
12. Aguilar Jr., "Rice in the Filipino Diet and Culture," 2–3.
13. O'Connor, "Agricultural Change and Ethnic Succession," 974–976.
14. Reid, *Southeast Asia in the Age of Commerce*, 1:20.
15. Allen, "Inland Angkor, Coastal Kedah," 79–87.
16. Stargardt, "From the Iron Age to Early Cities," 346.
17. Stargardt, "From the Iron Age to Early Cities," 350–351.
18. Higham, "At the Dawn of History," 433–434.
19. Higham, "At the Dawn of History," 424–425.
20. Hendrickson, "Transport Geographic Perspective on Travel and Communication," 449–450.
21. Higham, "At the Dawn of History," 425–426; Stargardt, "From the Iron Age to Early Cities," 350.
22. Bellwood, "Southeast Asia before History," 122–126.

23. Glover and Bellwood, *Southeast Asia*; Higham, "Mainland Southeast Asia from the Neolithic to the Iron Age," 58. Higham, *Archaeology of Mainland Southeast Asia*, 192–203.

24. Bellwood, "Southeast Asia before History," 123. The Han dynasty ruled between 202 BCE and 220 CE. Han also refers to the dominant ethnic group and culture that emerged in the Yellow River Valley region of China.

25. Bellwood, "Southeast Asia before History," 123.

26. Manguin, "Archaeology of Early Maritime Polities in Southeast Asia," 283–285.

27. Turnbull, *Mapping the World in the Mind*.

28. Turnbull, *Mapping the World in the Mind*, 14.

29. Shaffer, *Maritime Southeast Asia to 1500*.

30. Shaffer, *Maritime Southeast Asia to 1500*.

31. Manguin, "Southeast Asian Ship," 266–276.

32. Manguin, "Southeast Asian Ship," 267.

33. Manguin, "Southeast Asian Ship."

34. Manguin, "Southeast Asian Ship."

35. Manguin, "Archaeology of Early Maritime Polities," 283.

36. Manguin, "Southeast Asian Ship."

37. Shaffer, *Maritime Southeast Asia to 1500*; Manguin, "Archaeology of Early Maritime Polities."

38. Manguin, "Archaeology of Early Maritime Polities"; Hall, *Maritime Trade and State Development in Early Southeast Asia*, 20–21.

39. Wolters, *Early Indonesian Commerce*.

40. Manguin, "Archaeology of Early Maritime Polities."

41. Hall, *Maritime Trade and State Development in Early Southeast Asia*, 15–17.

42. Hall, *Maritime Trade and State Development in Early Southeast Asia*, 56.

43. Hall, *Maritime Trade and State Development in Early Southeast Asia*, 55.

44. Manguin, "Archaeology of Early Maritime Polities"; Hall, *Maritime Trade and State Development in Early Southeast Asia*, 55–56.

45. Manguin, "Archaeology of Early Maritime Polities," 300.

46. Manguin, "Archaeology of Early Maritime Polities," 300.

47. Wolters, *Early Indonesian Commerce*.

48. Wade, "Early Age of Commerce in Southeast Asia, 900–1300 CE," 221–265. Wade references Anthony Reid's use of the phrase to designate the period between the fourteenth and eighteenth centuries. In both cases, the authors emphasize the central role of changing trade relations in wider social and political change.

49. Wade, "Early Age of Commerce in Southeast Asia," 221–265.

50. Wade, "Early Age of Commerce in Southeast Asia," 232–233.

51. Wade, "Early Age of Commerce in Southeast Asia," 232–233.

52. Wade, "Early Age of Commerce in Southeast Asia," 248. The Kingdom of Mataram should not be confused with the Sultanate of Mataram, which was established in 1587. The Kingdom of Mataram is also known as the Medang Kingdom.

53. Wade, "Early Age of Commerce in Southeast Asia," 241. See also Li Tana, "Vietnam," 795–797.

54. Christie, "State Formation in Early Maritime Southeast Asia," 235–288.
55. Christie, "State Formation in Early Maritime Southeast Asia," 235–288.
56. Note that Bagan is also transliterated as Pagan.
57. Whitmore, "Rise of the Coast," 103–122.
58. Christie, "Javanese Markets and the Asian Sea Trade Boom," 352–353.
59. Lieberman, *Strange Parallels*, 21–44.
60. Lieberman, *Strange Parallels*, 91–92.
61. Hendrickson, "Transport Geographic Perspective on Travel and Communication," 444–457. See also Higham, *Civilization of Angkor*. The city of Angkor is also known as Yasodharapura.
62. Lieberman, *Strange Parallels*.
63. Koller, "Architectural Design at Bagan and Angkor," 93–141.
64. Lat, "Analysis of Construction Technologies in Pyu Cities and Bagan," 15–40.
65. Lat, "Analysis of Construction Technologies in Pyu Cities and Bagan."
66. Lat, "Analysis of Construction Technologies in Pyu Cities and Bagan."
67. Evans, Pottier, Fletcher, Hensley, Tapley, Milne, and Barbetti, "Comprehensive Archaeological Map," 14277–14282.
68. Wittfogel, *Oriental Despotism*.
69. Wolters, *History, Culture, and Region in Southeast Asian Perspectives*.
70. Van Setten van der Meer, *Sawah Cultivation in Ancient Java*.
71. Van Setten van der Meer, *Sawah Cultivation in Ancient Java*, 24.
72. Van Setten van der Meer, *Sawah Cultivation in Ancient Java*, 86.
73. Van Setten van der Meer, *Sawah Cultivation in Ancient Java*, 86.
74. Lansing, *Perfect Order*.
75. Lansing, *Perfect Order*.
76. Stark, "Pre-Angkorian and Angkorian Cambodia," 89–119.
77. O'Reilly and Shewan, "Phum Lovea," 468–483.
78. Hendrickson, "Transport Geographic Perspective on Travel and Communication"; Penny, Hall, Evans, and Polkinghorne, "Geoarchaeological Evidence," 4871–4876.
79. Penny et al., "Geoarchaeological Evidence"; Evans et al., "Comprehensive Archaeological Map."
80. Evans et al., "Comprehensive Archaeological Map."
81. Penny et al., "Demise of Angkor."
82. Penny et. al., "Geoarchaeological Evidence."
83. Lieberman, *Strange Parallels*, 92–97.
84. Lieberman, *Strange Parallels*, 92–97.
85. Lieberman, *Strange Parallels*, 92–97.

Chapter 3. Textiles, Commerce, and Sociotechnical Resilience

1. Much of Southeast Asian history is difficult to periodize in ways that make sense across the entire region. However, there are good reasons to adopt the term "early modern" without being regarded as Eurocentric. See Andaya, "Historicizing 'Modernity' in Southeast Asia," 391–409.
2. Reid and Cushman, *Sojourners and Settlers*.

3. Reid and Australian National University, "Seventeenth-Century Crisis in Southeast Asia," 639–659.

4. Reid and Australian National University, "Seventeenth-Century Crisis in Southeast Asia," 639–659. See also Reid, "Origins of Southeast Asian Poverty," 33–49.

5. Dijk, "VOC's Trade in Indian Textiles with Burma, 1634–80," 495–515; Reid, *Charting the Shape of Early Modern Southeast Asia*, 217–234; Hall, "Textile Industry of Southeast Asia, 1400–1800," 87–135.

6. Andaya, "Cloth Trade in Jambi and Palembang Society," 27–46, and *Flaming Womb*. See also Hall, "Textile Industry of Southeast Asia."

7. Hall, "Textile Industry of Southeast Asia," 100–101. For deeper technical and artistic detail on bark cloth and all other textiles described in this chapter, see Maxwell, *Textiles of Southeast Asia*.

8. Lieberman, *Strange Parallels*, 144.

9. Maxwell, *Textiles of Southeast Asia*, 162; Christie, "Texts and Textiles in 'Medieval' Java," 181–211, esp. 188.

10. Green, "Indic Impetus?," 277–313, esp. 303.

11. Maxwell, *Textiles of Southeast Asia*, 162.

12. Green, "Indic Impetus?," 303. Her main source for this was the memoir of Zhao Daguan from 1296.

13. For Java, see Christie, "Texts and Textiles in 'Medieval' Java," 188.

14. Lieberman, *Strange Parallels*, 144.

15. Andaya, "Cloth Trade in Jambi and Palembang Society," 32.

16. Christie, "Texts and Textiles in 'Medieval' Java."

17. Christie, "Texts and Textiles in 'Medieval' Java."

18. Christie, "Texts and Textiles in 'Medieval' Java," 186.

19. Christie, "Texts and Textiles in 'Medieval' Java," 186.

20. Hall mentions multiple designs of these so-called body-tension looms in "Textile Industry of Southeast Asia."

21. Maxwell, *Textiles of Southeast Asia*, 162.

22. Dijk, "VOC's Trade in Indian Textiles with Burma," 500.

23. Christie, "Texts and Textiles in 'Medieval' Java," 201. A contemporary dictionary of Bahasa Indonesia defines "cadar" as a veil. Stevens and Schmidgall-Tellings, *Comprehensive Indonesian-English Dictionary*.

24. Lieberman, *Strange Parallels*, 144–145; Green, "Indic Impetus?," 303.

25. Green, "Indic Impetus?"

26. Green, "Indic Impetus?," 192.

27. Green, "Indic Impetus?," 192. See also Christie, "Texts and Textiles in 'Medieval' Java."

28. Andaya, *Flaming Womb*, 115–117.

29. Kian, "End of the 'Age of Commerce'?," 292; Andaya, "Cloth Trade in Jambi and Palembang Society," and *Flaming Womb*.

30. Andaya, *Flaming Womb*. See also Reid, *Southeast Asia in the Age of Commerce*.

31. Note that few sources specify whether silk was reeled or woven in households. For that reason, I am not specifying here exactly what sort of fabric was being made,

although it is likely that the majority of it was cotton. Silk of some sort would have been a regulated luxury cloth, but it is likely that some was produced in households. See Andaya, *Flaming Womb*, for more context, esp. 115–120.

32. Andaya, *Flaming Womb*. Lack of documentation about such mundane activities makes it difficult to prove, however.

33. Andaya, *Flaming Womb*, 115–117. The literature and evidence on gender and textiles is somewhat biased toward the Malay world.

34. Andaya, *Flaming Womb*, 115–117.

35. Andaya, "Cloth Trade in Jambi and Palembang Society," 41. See also Christie, "Texts and Textiles in 'Medieval' Java" about privileges on Java for an earlier period.

36. Andaya, "Women and Economic Change," 165–190. Christie, "Texts and Textiles in 'Medieval' Java," 183. See also Dijk, "VOC's Trade in Indian Textiles with Burma."

37. Green, "Indic Impetus?," 277–313.

38. Hall, "Textile Industry of Southeast Asia."

39. Andaya, "Cloth Trade in Jambi and Palembang Society," 38. However, if sources allowed, it would be interesting to know more about how the types of weaving skills available waxed and waned across the region. See Green's discussion of Angkor's imports of fancy cloth in "Indic Impetus?"

40. See, for example, Dijk's discussion of the VOC's practices in Burma in "VOC's Trade in Indian Textiles with Burma."

41. Kian, "End of the 'Age of Commerce'?" 283–302, Reid, *Charting the Shape of Early Modern Southeast Asia*.

42. Andaya, "Cloth Trade in Jambi and Palembang Society," 40; Andaya, *Flaming Womb*, 115; Kian, "End of the 'Age of Commerce'?"

43. Kian, "End of the 'Age of Commerce,'?" 88.

44. Andaya, "Cloth Trade in Jambi and Palembang Society."

45. Andaya, "Cloth Trade in Jambi and Palembang Society." The evidence for this uptick in royal weaving is drawn both from the increased import of luxury thread and raw silk, and a change in practices of men's dress at the royal court that would have required the production of new clothing. The fabric was probably silk, but Andaya does not specify.

46. Andaya, "Cloth Trade in Jambi and Palembang Society."

47. Andaya, "Cloth Trade in Jambi and Palembang Society." Although Chinese traders had long-established trade in the region, after the fall of the Ming dynasty in 1644, many Ming loyalists left China to trade in Southeast Asia on a more permanent basis.

48. Andaya, "Cloth Trade in Jambi and Palembang Society," 40–41.

49. Andaya, "Cloth Trade in Jambi and Palembang Society."

50. Andaya, "Cloth Trade in Jambi and Palembang Society."

51. Andaya, "Cloth Trade in Jambi and Palembang Society."

52. Ricklefs, *History of Modern Indonesia since c. 1200*.

53. Kian, "Expansion of Chinese Inter-Insular and Hinterland Trade in Southeast Asia, c. 1400–1850."

Chapter 4. Localizing Foreign Technology

1. Giersch, *Asian Borderlands*.
2. Reid, "Chinese on the Mining Frontier in Southeast Asia," 21–36.
3. Reid, "Chinese on the Mining Frontier in Southeast Asia," 21–36.
4. Sun, "Shan Gems, Chinese Silver, and the Rise of Shan Principalities in Northern Burma."
5. Sun, "Shan Gems, Chinese Silver, and the Rise of Shan Principalities in Northern Burma."
6. Fiskesjö, "Mining, History, and the Anti-State Wa," 241–264.
7. Heidhues, *Golddiggers, Farmers, and Traders*, 51–52.
8. Heidhues, *Golddiggers, Farmers, and Traders*, 51–52.
9. Giersch, *Asian Borderlands*. It was also true in many parts of the Malay world as Chinese miners began to be brought in to exploit mines, and in the border areas between Vietnam and China. Heidhues, *Golddiggers, Farmers, and Traders*, and Vū, "Politics of Frontier Mining."
10. Vū, "Politics of Frontier Mining"; Nanny Kim, "Tangdan Copper Mines and the 1733 Earthquake," 51–74.
11. Fang et al., "Copper Mining and Smelting in Yunnan," 265–288.
12. Giersch, *Asian Borderlands*, 17–63.
13. Giersch, *Asian Borderlands*, 17–63.
14. Giersch, *Asian Borderlands*, 17–63. See also 1–16 in the same book.
15. Kim, "Tangdan Copper Mines and the 1733 Earthquake," 51–74.
16. Fang et al., "Copper Mining and Smelting in Yunnan." Also see Peter Golas, "Chemistry and Chemical Technology."
17. Fang et al., "Copper Mining and Smelting in Yunnan."
18. Fang et al., "Copper Mining and Smelting in Yunnan."
19. Fang et al., "Copper Mining and Smelting in Yunnan."
20. Fang et al., "Copper Mining and Smelting in Yunnan." For the Malay world, see Reid, "Chinese on the Mining Frontier in Southeast Asia."
21. Fang et al., "Copper Mining and Smelting in Yunnan."
22. Fang et al., "Copper Mining and Smelting in Yunnan."
23. Heidhues, "Chinese Settlements in Rural Southeast Asia," 172–177.
24. Heidhues, *Bangka Tin and Mentok Pepper*, 1–48.
25. Fiskesjö, "Mining, History, and the Anti-State Wa."
26. Giersch, *Asian Borderlands*.
27. For immigration in a later period, see Lees, *Planting Empire, Cultivating Subjects*.
28. McCarthy, "Yards at Cavite," 149–162; Li, "Ships and Shipbuilding in the Mekong Delta, c. 1750–1840."
29. McCarthy, "Yards at Cavite," 149–162.
30. McCarthy, "Yards at Cavite," 149–162.
31. McCarthy, "Yards at Cavite," 157.
32. McCarthy, "Yards at Cavite," 157.

33. McCarthy, "Yards at Cavite," 158.

34. Li, "Ships and Shipbuilding in the Mekong Delta," 119–138.

35. Li, "Ships and Shipbuilding in the Mekong Delta," 119–138.

36. Li, "Ships and Shipbuilding in the Mekong Delta," 119–138.

37. Li, "Ships and Shipbuilding in the Mekong Delta," 119–138.

38. Li, "Ships and Shipbuilding in the Mekong Delta," 119–138.

39. Li, "Ships and Shipbuilding in the Mekong Delta," 121–122.

40. Knaap, *Shallow Waters, Rising Tide*, 31–38.

41. Knaap and Sutherland, *Monsoon Traders*, 45.

42. Knaap and Sutherland, *Monsoon Traders*, 46.

43. Knaap and Sutherland, *Monsoon Traders*, 46.

Chapter 5. Intensification and Expansion

1. This area needs more investigation. The literature does not seem to have a great deal to say about cultivation practices in this transitional period, with some exceptions included here.

2. Giersch, *Asian Borderlands*.

3. The Tày ethnolinguistic group live in the Tai cultural area (see introduction) and are sometimes referred to as the Dai people.

4. Daniels, "Agricultural Technology and the Consolidation of Tay Polities in Northern Continental Southeast Asia during the 15th Century," 246–270.

5. Giersch, *Asian Borderlands*, 127–187.

6. Giersch, *Asian Borderlands*, 127–187.

7. Acabado, "Zones of Refuge," 180–195.

8. Acabado, "Zones of Refuge," 180–195.

9. Scott, *Art of Not Being Governed*.

10. Aguilar Jr., "Rice in Filipino Diet and Culture," 1–46.

11. Acabado, "Zones of Refuge."

12. Acabado, "Zones of Refuge."

13. Reid, "Chinese Trade and Southeast Asian Economic Expansion in the Later Eighteenth and Early Nineteenth Centuries," 21–34.

14. Lieberman, *Strange Parallels*, 297.

15. For Chinese sugar plantations, see Bosma, "Global Detour of Cane Sugar," 109–133.

16. Winn, "Slavery and Cultural Creativity on the Banda Islands," 365–389.

17. Bosma, "Global Detour of Cane Sugar."

18. Lieberman, *Strange Parallels*, 297.

19. Dove, "'Banana Tree at the Gate.'" See also Michael Dove's book-length historical anthropology that explores commodity production among the peoples of Borneo, *The Banana Tree at the Gate: A History of Marginal Peoples and Global Markets in Borneo*.

20. Dove, "'Banana Tree at the Gate.'"

21. Andaya, "Women and Economic Change."

22. Dove, "'Banana Tree at the Gate.'"

23. Dove, "'Banana Tree at the Gate.'"

24. Andaya, "Women and Economic Change."

25. Andaya, "Women and Economic Change."

26. Andaya, "Women and Economic Change."

27. Dove, "'Banana Tree at the Gate,'" 45–46; Andaya, "Women and Economic Change."

28. Dove, "'Banana Tree at the Gate,'" 45–46.

29. Trocki, "Origins of the Kangchu System, 1740–1860," 132–155. Although dealing with a slightly later period, Trocki's *Opium and Empire: Chinese Society in Colonial Singapore, 1800–1910* is also valuable for understanding Chinese plantation agriculture in the region.

Chapter 6. Technology and Cultures of Conflict in Early Modern Southeast Asia

1. Knaap, "Headhunting, Carnage and Armed Peace," 165–192; Sun, "Military Technology Transfers from Ming China," 495–517. A broad perspective on VOC warfare in the region can be found in Knaap, Den Heijer, and De Jong, *Oorlogen overzee*. Victor Lieberman's *Strange Parallels* provides a valuable overview of large-scale military conflict on the mainland.

2. Sen, "Intricacies of Premodern Asian Connections," 991–999.

3. Hacker, "Military Institutions, Weapons, and Social Change," 768–834.

4. Charney and Wellen, *Warring Societies of Pre-colonial Southeast Asia*, 14.

5. Charney, *Southeast Asian Warfare*.

6. Charney, *Southeast Asian Warfare*; Charney and Wellen, *Warring Societies of Pre-colonial Southeast Asia*.

7. Knaap, "Headhunting, Carnage and Armed Peace."

8. Charney, "Armed Rural Folk," 155–182.

9. Cipolla, *Guns and Sails in the Early Phase of European Expansion*.

10. Cipolla, *Guns and Sails in the Early Phase of European Expansion*.

11. For a nondeterministic approach to military history, see Black, *War and Technology*. Also see the illuminating special issue on technological determinism in military history in the journal *Vulcan* (vol. 7, no. 1), especially the introduction by journal editor Steven Walton: "Technological Determinism and the Study of War," 4–18.

12. Walton, "Technological Determinism and the Study of War"; Charney, *Southeast Asian Warfare*.

13. The classic work that makes this claim is Parker, *Military Revolution*.

14. Although somewhat dated, Barton C. Hacker offers a useful overview of the key issues in this debate in "Military Institutions, Weapons, and Social Change," 768–834. For an exploration of the revolution thesis in Asian history, see Lorge, *Asian Military Revolution*.

15. Charney, *Southeast Asian Warfare*.

16. Charney and Wellen, *Warring Societies of Pre-colonial Southeast Asia*, 1–19.

17. Charney and Wellen, *Warring Societies of Pre-colonial Southeast Asia*, 1–19.

18. For the mainland, see Lieberman, *Strange Parallels*.

19. Lieberman, *Strange Parallels*.

20. For examples of these justifications for warfare, see Charney, *Southeast Asian Warfare*; Pawakapan, "Warfare and Depopulation"; Andaya, "Aspects of Warfare in

Premodern Southeast Asia," 139–142; Andaya, "History, Headhunting and Gender in Monsoon Asia," 13–52; Hagerdal, "War and Culture," 81–118.

21. Pawakapan, "Warfare and Depopulation," 21.

22. Sun, "Military Technology Transfers from Ming China," 495–517; Whitmore, "Two Great Campaigns of the Hong Duc Era (1470–97) in Dai Viet," 119–136.

23. Sun, "Military Technology Transfers from Ming China."

24. Hawkley, "Reviving the Reconquista in Southeast Asia," 285–310. For an overview of Philippine history, see Nadeau, *History of the Philippines*.

25. Knaap, Den Heijer, and De Jong, *Oorlogen Overzee*.

26. Clulow and Mostert, *Dutch and English East India Companies*.

27. Andaya, "History, Headhunting and Gender in Monsoon Asia." Andaya is careful to point out that headhunting itself was a highly diverse practice with diverse cultural justifications (14–15). See also Knaap, "Headhunting, Carnage and Armed Peace in Amboina." Note that Maluku was previously referred to as the Moluccas in much European literature.

28. Knaap, "Headhunting, Carnage and Armed Peace in Amboina."

29. Andaya, "History, Headhunting and Gender in Monsoon Asia," 19–30.

30. Andaya, "History, Headhunting and Gender in Monsoon Asia," 19–30.

31. Andaya, "History, Headhunting and Gender in Monsoon Asia," 30–47.

32. Knaap, "Headhunting, Carnage, and Armed Peace in Amboina"; James Warren, "Tale of Two Centuries," 125–152; Wellen, "La Maddukelleng and Civil War in South Sulawesi," 47–72; Hagerdal, "War and Culture," and "Expansion and Internalization of Modes of Warfare on Pre-colonial Bali," 129–154.

33. Andaya, "History, Headhunting and Gender in Monsoon Asia," 21.

34. Andaya, "History, Headhunting and Gender in Monsoon Asia," 20.

35. Charney, *Southeast Asian Warfare*, 23–41.

36. Charney, *Southeast Asian Warfare*, 23–41. See also Hagerdal, "War and Culture."

37. Charney, *Southeast Asian Warfare*, 23–41.

38. Andaya, "Nature of War and Peace among the Bugis-Makassar People," 53–80.

39. Carey, "Political Economy of Poison," 517–543.

40. Carey, "Political Economy of Poison," 517–543.

41. Charney, *Southeast Asian Warfare*, 31–32.

42. Charney, *Southeast Asian Warfare*, 31–32.

43. Charney, *Southeast Asian Warfare*, 31–32. See also O'Connor, "Iron Working as Spiritual Inquiry in the Indonesian Archipelago," 173–190, and Frey, *Kris*.

44. Hagerdal, "War and Culture."

45. Andaya, "Nature of War and Peace," 57.

46. Charney, *Southeast Asian Warfare*, 63–64.

47. Andaya, "History, Headhunting and Gender in Monsoon Asia," 37.

48. Andaya, "History, Headhunting and Gender in Monsoon Asia," 37.

49. For the Bugis, see Andaya, "Nature of War and Peace," 55, and Andaya, "History, Headhunting and Gender in Monsoon Asia," 42.

50. Andaya, "History, Headhunting and Gender in Monsoon Asia," 42.

51. Warren, "Tale of Two Centuries." See also Warren's *Sulu Zone, 1768–1898*.

52. Warren, "Tale of Two Centuries," 134.

53. Pawakapan, "Warfare and Depopulation."

54. Charney, *Southeast Asian Warfare*, 42–72; Warren, "Tale of Two Centuries"; Pawakapan, "Warfare and Depopulation."

Chapter 7. Scaling Up Warfare

1. Charney and Wellen, *Warring Societies of Pre-colonial Southeast Asia*, 1–20.

2. Lieberman, "Some Comparative Thoughts on Premodern Southeast Asian Warfare," 215–225.

3. Knaap, "Headhunting, Carnage and Armed Peace."

4. Knaap, "Headhunting, Carnage and Armed Peace," 174–181.

5. Hagerdal, "Expansion and Internalization of Modes of Warfare."

6. Charney, "Armed Rural Folk."

7. Dutton, "Flaming Tiger, Burning Dragon," 48–93.

8. Andaya, "Nature of War and Peace among the Bugis-Makassar People," 3–80.

9. Hagerdal, "War and Culture," and "Expansion and Internalization of Modes of Warfare."

10. Warren, "Tale of Two Centuries."

11. Charney argues that sources for Burma are also of high quality. It is possible that more research on the Iranun would reveal a history of intensive transformation as well.

12. Sun, "Military Technology Transfers."

13. Sun, "Military Technology Transfers."

14. Whitmore, "Two Great Campaigns."

15. Sun, "Military Technology Transfers"; see also Whitmore, "Two Great Campaigns."

16. Sun, "Military Technology Transfers."

17. Sun, "Military Technology Transfers."

18. Sun, "Military Technology Transfers."

19. Dutton, "Flaming Tiger, Burning Dragon," 66.

20. Dutton, "Flaming Tiger, Burning Dragon," 66.

21. Dutton, "Flaming Tiger, Burning Dragon," 60–71. Dutton does not provide the name of the original author of this manual.

22. Vu, "Age of Sea Falcons," 115.

23. Vu, "Age of Sea Falcons," 110.

24. Hagerdal, "Expansion and Internalization of Modes of Warfare."

25. Hagerdal, "Expansion and Internalization of Modes of Warfare," 136.

26. Warren, "Tale of Two Centuries." See also Hagerdal, "Expansion and Internalization of Modes of Warfare."

27. Pawakapan, "Warfare and Depopulation," 23.

28. Beemer, "Southeast Asian Slavery and Slave-Gathering Warfare," 481–506. See also Charney, *Southeast Asian Warfare*, 42–72.

29. Charney, *Southeast Asian Warfare*, 48.

30. Charney, *Southeast Asian Warfare*, 42–72.

31. Charney, "Crisis and Reformation," 185–219.

32. Pawakapan, "Warfare and Depopulation."

33. Lieberman, "Some Comparative Thoughts."

34. Lieberman, *Strange Parallels.*

35. Sun, "Military Technology Transfers."

36. Sun, "Military Technology Transfers." See also Hagerdal, "War and Culture"; Charney, *Southeast Asian Warfare*, and "Shallow-Draft Boats, Guns, and the Aye-Ra-Wa-Ti," 16–63; Warren, "Tale of Two Centuries"; and Dutton, "Flaming Tiger, Burning Dragon."

37. Sun, "Military Technology Transfer," 496–497.

38. Charney, "Shallow-Draft Boats"; Charney, *Southeast Asian Warfare*, 42–72, 104–130.

39. Carey, "Political Economy of Poison."

40. Rodriguez, "Juan de Salcedo Joins the Native Form of Warfare"; Knaap, "Headhunting, Carnage and Armed Peace."

41. Charney, *Southeast Asian Warfare.*

42. Charney, *Southeast Asian Warfare*, 42–72.

43. Charney, *Southeast Asian Warfare*, 42–72. See also Andaya, "Nature of War and Peace," and Hagerdal, "Expansion and Internalization of Modes of Warfare."

44. Hagerdal, "Expansion and Internalization of Modes of Warfare," 152.

45. Andaya, "Nature of War and Peace," and Fiskesjö, "Mining, History, and the Anti-State Wa," 257.

46. Charney, "Crisis and Reformation."

47. Knaap, "Headhunting, Carnage and Armed Peace."

48. Charney, *Southeast Asian Warfare*, 53.

49. H. H. Kang explores this subject for Chosŏn Korea in "Crafting Knowledge." It offers an interesting model for Southeast Asianists to explore.

50. Souza, "Albuquerque, Afonso de, 1453–1515."

51. Andrade, "Was the European Sailing Ship a Key Technology of European Expansion?" 17–40.

52. Andrade, "Beyond Guns, Germs, and Steel," 165–186.

53. Vu, "Age of Sea Falcons," 112.

54. Charney, "Shallow-Draft Boats."

55. Charney, "Shallow-Draft Boats"; Vu "Age of Sea Falcons."

56. I am grateful to Larrie Ferreiro for his insight into the challenges of adding guns to ships not originally designed to carry them.

57. Knaap, "Headhunting, Carnage and Armed Peace."

58. Sen, "Intricacies of Premodern Asian Connections," 991–999.

Chapter 8. Technology, Empire, and Nationalism

1. See discussions in Elson, *Javanese Peasants and the Colonial Sugar Industry*; Coclanis, "Metamorphosis," 35–67; Bosma, *Making of a Periphery.*

2. Relevant book-length treatments include Aso, *Rubber and the Making of Vietnam*; Anderson, *Colonial Pathologies*; Biggs, *Quagmire*; Adas, *Burma Delta*; Mrázek, *Engineers of Happy Land*; and Moon, *Technology and Ethical Idealism.*

3. Clapperton Chakanetse Mavhunga makes this point in compelling fashion for Africa, in *Transient Workspaces*.

4. Daniel Headrick popularized this approach in *Tools of Empire*.

5. Chatterjee, *Politics of the Governed*.

6. See, for example, Stoler, *Capitalism and Confrontation in Sumatra's Plantation Belt, 1870–1979*; Aso, *Rubber and the Making of Vietnam*; Bosma, *Making of a Periphery*.

7. Adas, *Burma Delta*, 15–82.

8. Coclanis, "Metamorphosis."

9. Adas, *Burma Delta*, 97.

10. Adas, *Burma Delta*, 31, 127.

11. Adas, *Burma Delta*, 128–131.

12. Coclanis, "Metamorphosis."

13. Mintz, *Sweetness and Power*.

14. Aguilar Jr., "Sugar Planter-State Relations," 50–80.

15. Aguilar Jr., "Sugar Planter-State Relations," 50–80.

16. This was no small feat. See, for comparison, Otremba, "Inventing Ingenios," 119–147.

17. Aguilar Jr., "Sugar Planter-State Relations."

18. Storey, *Science and Power in Colonial Mauritius*.

19. Follett et al., *Plantation Kingdom*.

20. Aguilar, "Fulcrum of Structure-Agency," 87–122.

21. Ricklefs, *History of Modern Indonesia*, 144–157.

22. Elson, *Javanese Peasants and the Colonial Sugar Industry*.

23. Elson, *Javanese Peasants and the Colonial Sugar Industry*, 71–97.

24. Elson, *Javanese Peasants and the Colonial Sugar Industry*, 71–97.

25. Bosma and Curry-Machado, "Two Islands, One Commodity," 237–262.

26. Moon, *Technology and Ethical Idealism*, 109–124.

27. Moon, *Technology and Ethical Idealism*, 109–124.

28. Aso, *Rubber and the Making of Vietnam*.

29. Aso, "The Scientist, the Governor, and the Planter," 231–256.

30. Aso, "The Scientist, the Governor, and the Planter," 231–256. Elsewhere, smallholder rubber was extremely successful, especially in Indonesia.

31. Dove, *Banana Tree at the Gate*.

32. Tagliacozzo, "Necklace of Fins," 23–48.

33. Tagliacozzo, "Necklace of Fins," 23–48.

34. Bryant, "Shifting the Cultivator," 225–250.

35. Bryant, "Shifting the Cultivator," 225–250.

36. Godfrey, *Submarine Telegraphy and the Hunt for Gutta Percha*, 60–95.

37. Godfrey, *Submarine Telegraphy and the Hunt for Gutta Percha*, 89–92.

38. Godfrey, *Submarine Telegraphy and the Hunt for Gutta Percha*, 95–134.

39. Dove, *Banana Tree at the Gate*, 78–96.

40. Dove, *Banana Tree at the Gate*, 75.

41. Godfrey, *Submarine Telegraphy and the Hunt for Gutta Percha*, 212–213.

42. Godfrey, *Submarine Telegraphy and the Hunt for Gutta Percha*, 212–213.

43. *American Heritage Dictionary of the English Language*, 5th edition.

44. Star, "Ethnography of Infrastructure," 377–391.

45. See, for example, Adas, *Burma Delta*.

46. Ouyyanont, "Physical and Economic Change in Bangkok," 437–474.

47. Ouyyanont, "Physical and Economic Change in Bangkok," 437–474.

48. Biggs, *Quagmire*.

49. Biggs, *Quagmire*, 23–52.

50. Biggs, *Quagmire*, 23–52.

51. Biggs, *Quagmire*, 23–52.

52. Jasanoff and Kim, *Dreamscapes of Modernity*.

53. Ross, "Tin Frontier," 454–479.

54. Mrázek, *Engineers of Happyland*, 4.

55. Kaur, " 'Hantu' and Highway," 1–49

56. Kaur, " 'Hantu' and Highway," 1–49.

57. The Iban people were uninterested in this form of labor, as demonstrated by earlier experiences in telegraph building.

58. Kaur, " 'Hantu' and Highway."

59. Rousseau, "Imperial Railway Failure," 1–17.

60. Holm, "Role of State Railways in Thai History, 1892–1932," 34–105.

61. Holm, "Role of State Railways in Thai History, 1892–1932," 61–62.

62. Rousseau, "Imperial Railway Failure."

63. Holm, "Role of State Railways in Thai History, 1892–1932."

64. Holm, "Role of State Railways in Thai History, 1892–1932," 34–105.

65. Holm, "Role of State Railways in Thai History, 1892–1932," 34–105.

66. Holm, "Role of State Railways in Thai History, 1892–1932," 110–113.

67. Holm, "Role of State Railways in Thai History, 1892–1932," 110–113.

68. Mrazek, *Engineers of Happyland*, 91.

69. Mrazek, *Engineers of Happyland*, 8.

70. Holm, "Role of State Railways in Thai History, 1892–1932," 55–56.

71. Holm, "Role of State Railways in Thai History, 1892–1932," 99.

72. Del Testa, " 'Imperial Corridor,' " 372; Mrazek, *Engineers of Happyland*.

73. Del Testa, " 'Imperial Corridor.' "

74. Del Testa, " 'Imperial Corridor.' "

75. Mrazek, *Engineers of Happyland*, 14.

76. Del Testa, " 'Imperial Corridor.' "

77. Shiraishi, *Age in Motion*.

78. Del Testa, " 'Imperial Corridor.' "

79. Winichakul, "The Quest for 'Siwilai,' " 528–549.

80. Moon, *Technology and Ethical Idealism*.

Acabado, Stephen. "Zones of Refuge: Resisting Conquest in the Northern Philippine Highlands through Environmental Practice." *Journal of Anthropological Archaeology* 52 (2018): 180–195.

Adas, Michael. *The Burma Delta: Economic Development and Social Change on an Asian Rice Frontier, 1852–1941.* Madison: University of Wisconsin Press, 1974.

Aguilar, Filomeno V., Jr. "The Fulcrum of Structure-Agency: History and Sociology of Sugar Haciendas in Colonial Negros." *Sociological Review* 61, no. 1 (2013): 87–122.

———. "Rice in the Filipino Diet and Culture." Research Paper Series, Philippine Institute for Development Studies, July 2005, 1–52.

———. "Sugar Planter-State Relations and Labour Processes in Colonial Philippine Haciendas." *Journal of Peasant Studies* 22, no. 1 (1994): 50–80.

Ali, Daud. "The Early Inscriptions of Indonesia and the Problems of the Sanskrit Cosmopolis." In *Early Interactions between South and Southeast Asia*, edited by Pierre Yves Manguin, A. Mani, and Geoff Wade, 276–297. Singapore: ISEAS, 2011.

Allen, Jane. "Inland Angkor, Coastal Kedah: Landscapes, Subsistence Systems and State Development in Early Southeast Asia." *Bulletin of the Indo-Pacific Prehistory Association* 16 (1997): 79–87.

Amos, Jean-Claude. "Les bateaux vietnamiens dessinés par François-Edmond Paris." *Techniques and Culture* 35–36 (2001): 1–6.

Andaya, Barbara Watson. "Aspects of Warfare in Premodern Southeast Asia." *Journal of the Economic and Social History of the Orient* 46, no. 2 (2003): 139–142.

———. "The Cloth Trade in Jambi and Palembang Society during the Seventeenth and Eighteenth Centuries." *Indonesia* 48 (October 1989): 27–46.

———. *The Flaming Womb: Repositioning Women in Early Modern Southeast Asia.* Honolulu: University of Hawai'i Press, 2006.

———. "Historicizing 'Modernity' in Southeast Asia." *Journal of the Economic and Social History of the Orient* 40, no. 4 (1997): 391–409.

———. "History, Headhunting and Gender in Monsoon Asia: Comparative and Longitudinal Views." *South East Asia Research* 12, no. 1 (2004): 13–52.

———. "Women and Economic Change: The Pepper Trade in Pre-Modern Southeast Asia." *Journal of Economic and Social History of the Orient* 38 (1995): 165–190.

Andaya, Leonard. Y. "Nature of War and Peace among the Bugis-Makassar People." *South East Asia Research* 12, no. 1 (2004): 53–80.

Anderson, Warwick. *Colonial Pathologies: American Tropical Medicine, Race, and Hygiene in the Philippines*. Durham, NC: Duke University Press, 2006.

Andrade, Tonio. "Beyond Guns, Germs, and Steel: European Expansion and Maritime Asia 1400–1750." *Journal of Early Modern History* 14 (2010): 165–186.

———. "Was the European Sailing Ship a Key Technology of European Expansion?" *International Journal of Maritime History* 23, no. 2 (2011): 17–40.

Aso, Michitake. *Rubber and the Making of Vietnam: An Ecological History, 1897–1975*. Chapel Hill: University of North Carolina Press, 2018.

———. "The Scientist, the Governor, and the Planter: The Political Economy of Agricultural Knowledge in Indochina during the Creation of a 'Science of Rubber,' 1900–1940." *East Asian Science, Technology and Society* 3, no. 2 (2009): 231–256.

Association of Southeast Asian Nations. "ASEAN Statistical Yearbook, 2020." Jakarta: ASEAN Secretariat, December 2020.

Baer, A. S. "The Importance of Tools in Orang Asli Prehistory." In *Malaysia's Original People: Past, Present, and Future of the Orang Asli*, edited by Kirk Endicott, 153–172. Singapore: NUS Press, 2016.

Beemer, Bryce. "Southeast Asian Slavery and Slave-Gathering Warfare as a Vector for Cultural Transmission: The Case of Burma and Thailand." *Historian* 71, no. 3 (2009): 481–506.

Bellina, Bérénice, and Ian Glover. "The Archaeology of Early Contact with India and the Mediterranean World from the Fourth Century BC to the Fourth Century AD." In *Southeast Asia: From Prehistory to History*, edited by Ian Glover and Peter Bellwood, 68–89. London: Routledge Curzon, 2004.

Bellwood, Peter S. *First Farmers: The Origins of Agricultural Societies*. Malden, MA: Blackwell, 2005.

———. *First Islanders: Prehistory and Human Migration in Island Southeast Asia*. Oxford: Wiley Blackwell, 2017.

———. "Southeast Asia Before History: From Prehistory to c. 1500." In *Cambridge History of Southeast Asia*, vol. 1, edited by Nicholas Tarling, 55–136. Cambridge: Cambridge University Press, 1992.

Biggs, David A. *Quagmire: Nation-Building and Nature in the Mekong Delta*. Seattle: University of Washington Press, 2010.

Black, Jeremy. *War and Technology*. Bloomington: Indiana University Press, 2013.

Bosma, Ulbe. "The Global Detour of Cane Sugar." In *Colonialism, Institutional Change, and Shifts in Global Labor Relations*, edited by Karin Hofmeester and Pim De Zwart. Amsterdam: Amsterdam University Press, 2018.

———. *The Making of a Periphery: How Island Southeast Asia Became a Mass Exporter of Labor*. New York: Columbia University Press, 2019.

Bosma, Ulbe, and Jonathan Curry-Machado. "Two Islands, One Commodity: Cuba, Java, and the Global Sugar Trade (1790–1930)." *New West Indian Guide* 86, no. 34 (2012): 237–262.

Bray, Francesca. *The Rice Economies: Technology and Development in Asian Societies*. Berkeley: University of California Press, 1994.

Bryant, Raymond L. "Shifting the Cultivator: The Politics of Teak Regeneration in Colonial Burma." *Modern Asian Studies* 28, no. 2 (1994): 225–250.

Carey, Daniel. "The Political Economy of Poison: The Kingdom of Makassar and the Early Royal Society." *Renaissance Studies* 17, no. 3 (2003): 517–543.

Castillo, Cristina. "Rice in Thailand: The Archaeobotanical Contribution." *Rice* 4 (2011): 114–120.

Charney, Michael W. "Armed Rural Folk: Elements of Precolonial Warfare in the Artistic Representations and Written Accounts of the Pacification Campaign (1886–1889) in Burma." In Charney and Wellen, *Warring Societies of Pre-colonial Southeast Asia*, 155–182.

———. "Crisis and Reformation in a Maritime Kingdom of Southeast Asia: Forces of Instability and Political Disintegration in Western Burma (Arakan), 1603–1701." *Journal of the Economic and Social History of the Orient* 41, no. 2 (2015): 1851219.

———. "Shallow-Draft Boats, Guns, and the Aye-Ra-Wa-Ti: Continuity and Change in Ship Structure and River Warfare in Precolonial Myanma." *Oriens Extremus* 40, no. 1 (1997): 16–63.

———. *Southeast Asian Warfare, 1300–1900*. Leiden: Brill, 2004.

Charney, Michael W., and Kathryn Wellen, eds. *Warring Societies of Pre-colonial Southeast Asia: Local Cultures of Conflict within a Regional Context*. Copenhagen: NIAS Press, 2018.

Chatterjee, Partha. *The Politics of the Governed Reflections on Popular Politics in Most of the World*. New York: Columbia University Press, 2004.

Christie, Jan Wisseman. "Javanese Markets and the Asian Sea Trade Boom of the Tenth to Thirteenth Centuries A.D." *Journal of the Economic and Social History of the Orient* 41, no. 3 (1998): 344–381.

———. "State Formation in Early Maritime Southeast Asia: A Consideration of the Theories and the Data." *Bijdragen tot de Taal-, Land- en Volkenkunde* 151 (1995): 235–288.

———. "Texts and Textiles in 'Medieval' Java." *Bulletin de l'Ecole française d'Extrême-Orient* 80 (1993): 181–211.

Cipolla, C. M. *Guns and Sails in the Early Phase of European Expansion, 1400–1700*. London: Collins, 1965.

Clulow, Adam, and Tristan Mostert, eds. *The Dutch and English East India Companies: Diplomacy, Trade, and Violence in Early Modern Asia*. Amsterdam: Amsterdam University Press, 2018.

Coclanis, Peter A. "Metamorphosis: The Rice Boom, Environmental Transformation, and the Problem of Truncation in Colonial Lower Burma, 1850–1940." *Agricultural History* 93, no. 1 (2019): 35–67.

Cooke, Nola, and Li Tana. *Water Frontier: Commerce and the Chinese in the Lower Mekong Region, 1750–1880*. Singapore: Singapore University Press, 2004.

Dallos, Csilla. "Beyond Economic Gain: Strategic Use of Trade in Boundary Maintenance by Semang Collectors of Peninsular Malaysia." In *Malaysia's Original People: Past, Present and Future*, edited by Kirk Endicott, 403–422. Singapore: National University of Singapore Press, 2016.

Daniels, Christian. "Agricultural Technology and the Consolidation of Tay Polities in Northern Continental Southeast Asia during the 15th Century." In *Southeast Asia in the Fifteenth Century: The China Factor*, edited by Geoff Wade and Sun Laichen, 246–270. Singapore: National University of Singapore Press, 2010.

Del Testa, David W. "'Imperial Corridor': Association, Transportation and Power in French Colonial Indochina." *Science, Technology and Society* 4, no. 2 (1999): 372.

Dijk, Wil O. "The VOC's Trade in Indian Textiles with Burma, 1634–80." *Journal of Southeast Asian Studies* 33, no. 3 (2002): 495–515.

Diller, Anthony. "Tai-Kadai Languages." In *Encyclopedia of Modern Asia*, vol. 5, edited by Karen Christensen and David Levinson, 370–372. New York: Charles Scribner's Sons, 2002.

Dove, Michael. *The Banana Tree at the Gate: A History of Marginal Peoples and Global Markets in Borneo.* New Haven, CT: Yale University Press, 2011.

———. "The 'Banana Tree at the Gate': Perceptions of the Production of *Piper Nigrum* (Piperaceae) in a Seventeenth Century Malay State." *Economic Botany* 51, no. 4 (1997): 347–361.

Dutton, George. "Flaming Tiger, Burning Dragon: Elements of Early Modern Vietnamese Military Technology." *East Asian Science, Technology, and Medicine* 21 (2003): 48–93.

Edgerton, David. *The Shock of the Old: Technology and Global History since 1900.* Oxford: Oxford University Press, 2007.

Elson, R. E. *Javanese Peasants and the Colonial Sugar Industry: Impact and Change in an East Java Residency, 1830–1940.* Singapore: Oxford University Press, 1984.

Enfield, N. J. "Linguistic Diversity in Mainland Southeast Asia." In *Dynamics of Human Diversity*, edited by N. J. Enfield, 63–79. Canberra: Pacific Linguistics, 2011.

Evans, Caitlin, Nigel Chang, and Naho Shimizu. "Sites, Survey, and Ceramics: Settlement Patterns of the First to Ninth Centuries CE in the Upper Mun River Valley, Northeast Thailand." *Journal of Southeast Asian Studies* 47, no. 3 (October 2016): 438–467.

Evans, Damian, Christophe Pottier, Roland Fletcher, Scott Hensley, Ian Tapley, Anthony Milne, and Michael Barbetti. "A Comprehensive Archaeological Map of the World's Largest Preindustrial Settlement Complex at Angkor, Cambodia." *Proceedings of the National Academy of the Sciences of the United States* 104, no. 36 (September 4, 2007): 14277–14282.

Fang Zhuofen, Hu Tiewen, Jian Rui, and Fang Xing. "Copper Mining and Smelting in Yunnan." In *Chinese Capitalism, 1522–1840*, edited by Xu Dixin and Wu Chenming, 265–288. New York: St. Martin's Press, 2000.

Fiskesjö, Magnus. "Mining, History, and the Anti-State Wa: The Politics of Autonomy between Burma and China." *Journal of Global History* 5, no. 2 (2010): 241–264.

Foley, William A. *The Papuan Languages of New Guinea.* Cambridge: Cambridge University Press 1986.

Follett, Richard, Sven Beckert, Peter A. Coclanis, and Barbara Hahn. *Plantation Kingdom: The American South and Its Global Commodities.* Baltimore: Johns Hopkins University Press, 2016.

Food and Agriculture Organization of the United Nations. FAOSTAT. http://www.fao
.org/faostat/en/#data/QC.

Frey, Edward. *Kris: Mystic Weapon of the Malay World*. Oxford: Oxford University
Press, 2003.

Georg, Stefan. "Hmong-Mien Languages." In *Encyclopedia of Modern Asia*, vol. 5,
edited by Karen Christensen and David Levinson, 535–546. New York: Charles
Scribner's Sons, 2002.

———. "Sino-Tibetan Languages." In *Encyclopedia of Modern Asia*, vol. 5, edited by
Karen Christensen and David Levinson, 234–236. New York: Charles Scribner's
Sons, 2002.

Giersch, C. Patterson. *Asian Borderlands: The Transformation of Qing China's Yunnan
Frontier*. Cambridge, MA: Harvard University Press, 2006.

Gittinger, Mattiebelle. *Splendid Symbols: Textiles and Tradition in Indonesia*. Washington, DC: Textile Museum, 1979.

Glover, Ian, and Peter S. Bellwood. *Southeast Asia: From Prehistory to History*. London:
Routledge Curzon, 2004.

Godfrey, Helen. *Submarine Telegraphy and the Hunt for Gutta Percha*. Leiden: Brill,
2018.

Golas, Peter. "Chemistry and Chemical Technology." In *Science and Civilization in
China*, vol. 5, part XIII, *Mining*, edited by Joseph Needham. Cambridge: Cambridge University Press, 1999.

Green, Gillian. "Indic Impetus? Innovations in Textile Usage in Angkorian Period
Cambodia." *Journal of the Economic and Social History of the Orient* 43, no. 3
(2000): 277–313.

Großmann, Kristina. "'Dayak, Wake Up': Land, Indigeneity, and Conflicting Ecologies
in Central Kalimantan, Indonesia." *Bijdragen Tot De Taal-, Land- En Volkenkunde* 175, no. 1 (2019): 1–28.

Ha, Van Tan. "The Hoabinhian and Before." *Bulletin of the Indo-Pacific Prehistory
Association* 16 (1997): 35–41.

Hacker, Barton C. "Military Institutions, Weapons, and Social Change: Toward a New
History of Military Technology." *Technology and Culture* 35, no. 4 (1994): 768–834.

Hagerdal, Hans. "Expansion and Internalization of Modes of Warfare in Pre-colonial
Bali." In *Warring Societies of Pre-Colonial Southeast Asia: Local Cultures of
Conflict within a Regional Context*, edited by Michael W. Charney and Kathryn
Wellen, 129–154. Copenhagen: Nordic Institute of Asian Studies, 2018.

———. "War and Culture: Balinese and Sasak Views on Warfare in Traditional Historiography." *South East Asia Research* 12, no. 1 (2018): 81–118.

Hall, Kenneth R. *Maritime Trade and State Development in Early Southeast Asia*.
Honolulu: University of Hawai'i Press, 1985.

———. "The Textile Industry of Southeast Asia, 1400–1800." *Journal of the Economic
and Social History of the Orient* 39, no. 2 (1996): 87–135.

Hawkley, Ethan P. "Reviving the Reconquista in Southeast Asia: Moros and the Making of the Philippines, 1565–1662." *Journal of World History* 25, no. 2/3 (June/
September 2014): 285–310.

Headrick, Daniel. *Tools of Empire: Technology and European Imperialism in the Nineteenth Century.* Oxford: Oxford University Press, 1981.

Heidhues, Mary Somers. *Bangka Tin and Mentok Pepper: Chinese Settlement on an Indonesian Island.* Singapore: Institute of Southeast Asian Studies, 1992.

———. "Chinese Settlements in Rural Southeast Asia: Unwritten Histories." In *Sojourners and Settlers: Histories of Southeast Asia and the Chinese*, edited by Anthony Reid, 164–182. Honolulu: University of Hawai'i Press, 1996.

———. *Golddiggers, Farmers, and Traders in the "Chinese District" of West Kalimantan, Indonesia.* Ithaca, NY: Cornell University Press, 2003.

Hendrickson, Mitch. "A Transport Geographic Perspective on Travel and Communication in Angkorian Southeast Asia (Ninth to Fifteenth Centuries AD)." *World Archaeology* 43, no. 3 (2011): 444–457.

Higham, Charles. *The Archaeology of Mainland Southeast Asia: From 10,000 B.C. to the Fall of Angkor Cambridge World Archaeology.* Cambridge: Cambridge University Press, 1989.

———. "At the Dawn of History: From Iron Age Aggrandisers to Zhenla Kings." *Journal of Southeast Asian Studies* 47, no. 3 (October 2016): 418–437.

———. *The Civilization of Angkor.* London: Weidenfeld and Nicolson, 2001.

———. "Hunter-Gatherers in Southeast Asia: From Prehistory to the Present." *Human Biology* 85, nos. 1–3 (2013): 21–43.

———. "Mainland Southeast Asia from the Neolithic to the Iron Age." In *Southeast Asia: From Prehistory to History*, edited by Ian Glover and Peter Bellwood, 41–67. London: Routledge Curzon, 2004.

Holm, David Frederick. "The Role of State Railways in Thai History, 1892–1932." PhD diss., Yale University, 1978.

Jasanoff, Sheila, and Sang-Hyun Kim, eds. *Dreamscapes of Modernity: Sociotechnical Imaginaries and the Fabrication of Power.* Chicago: University of Chicago Press, 2015.

Jepsen, Martin Rudbeck, Matilda Palm, and Thilde Bech Bruun. "What Awaits Myanmar's Uplands Farmers? Lessons Learned from Mainland Southeast Asia." *Land* 8, no. 2 (2019): 1–23.

Junker, Laura Lee. "Hunter-Gatherer Landscapes and Lowland Trade in the Prehispanic Philippines." *World Archaeology* 27, no. 3 (1996): 389–410.

———. *Raiding, Trading, and Feasting: The Political Economy of Philippine Chiefdoms.* Honolulu: University of Hawai'i Press, 1999.

Kang, H. H. "Crafting Knowledge: Artisan, Officer, and the Culture of Making in Chosŏn Korea, 1392–1910." PhD diss., Harvard University, 2020.

Kaur, Amarjit. "'Hantu' and Highway: Transport in Sabah, 1881–1963." *Modern Asian Studies* 28 (1994): 1–49.

Kian, Kwee Hui. "The End of the 'Age of Commerce'? Javanese Cotton Trade Industry from the Seventeenth to the Eighteenth Centuries." In *Chinese Circulations: Capital, Commodities, and Networks in Southeast Asia*, edited by Eric Tagliacozzo and Weng-Chin Chang, 283–302. Durham, NC: Duke University Press, 2011.

———. "The Expansion of Chinese Inter-Insular and Hinterland Trade in Southeast Asia, c. 1400–1850." In *Environment, Trade and Society in Southeast Asia: A Longue Durée Perspective*, edited by David Henley and Henk Schulte Nordholt, 149–165. Leiden: Brill, 2015.

Kieschnick, John. *The Impact of Buddhism on Chinese Material Culture*. Princeton, NJ: Princeton University Press, 2003.

Kim, Nanny. "The Tangdan Copper Mines and the 1733 Earthquake: A Mining Community before the Boom in the Far Southwest of Qing China." *Cross-Currents: East Asian History and Culture Review*, no. 13 (December 2014): 51–74.

Knaap, Gerrit. "Headhunting, Carnage and Armed Peace in Amboina, 1500–1700." *Journal of the Economic and Social History of the Orient* 46, no. 2 (2003): 165–192.

———. *Shallow Waters, Rising Tide: Shipping and Trade in Java around 1775*. Leiden: KITLV Press, 1996.

Knaap, Gerrit, Henk den Heijer, and Michiel de Jong. *Oorlogen Overzee: Militair Optreden door Compagnie en Staat buiten Europa, 1595–1814*. Amsterdam: Boom, 2015.

Knaap, Gerrit, and Heather Sutherland. *Monsoon Traders: Ships, Skippers and Commodities in Eighteenth-Century Makassar*. Leiden: KITLV Press, 2004.

Koller, Alexander. "Architectural Design at Bagan and Angkor: A Comparison." *Journal of the Royal Asiatic Society* 27, no. 1: 1–49.

Kuchikura, Yukio. "Efficiency and Focus of Blowpipe Hunting among Semaq Beri Hunter-Gatherers of Peninsular Malaysia." *Human Ecology* 16, no. 3 (1988): 271–305.

Lansing, John Stephen. "The 'Indianization' of Bali." *Journal of Southeast Asian Studies* 14, no. 2 (1983): 409–421.

———. *Perfect Order: Recognizing Complexity in Bali Princeton Studies in Complexity*. Princeton, NJ: Princeton University Press, 2006.

Lat, Kyaw. "Analysis of Construction Technologies in Pyu Cities and Bagan." In *Bagan and the World: Early Myanmar and Its Global Connections*, edited by Goh Geok Yian, John N. Miksic, and Michael Aung-Thwin, 15–40. Singapore: ISEAS Publishing, 2018.

Lees, Lynn Hollen. *Planting Empire, Cultivating Subjects: British Malaya, 1786–1941*. Cambridge: Cambridge University Press, 2017.

Lemke, Ashley. "Hunter-Gatherers and Archaeology." In *Foraging in the Past: Archaeological Studies of Hunter-Gatherer Diversity*, 3–18. Boulder: University of Colorado Press, 2018.

Li, Tana. "Ships and Shipbuilding in the Mekong Delta, c. 1750–1840." In *Water Frontier: Commerce and the Chinese in the Lower Mekong Region, 1750–1880*. Edited by Nola Cooke and Li Tana, 119–135. Singapore: Rowman and Littlefield, 2004.

———. "Vietnam." In *History of World Trade since 1450*, vol. 2, edited by John J. McCusker, 795–797. Detroit: Macmillan Reference USA, 2006.

Lieberman, Victor. "Some Comparative Thoughts on Premodern Southeast Asian Warfare." *Journal of the Economic and Social History of the Orient* 46, no. 2 (2003): 215–225.

———. *Strange Parallels: Integration on the Mainland*. Cambridge: Cambridge University Press, 2003.

L[odewijcksz], W[illem]. Prima Pars Descriptionis Iteneris Navalis in Indian Orientalem, Earvmqve Rervm Qv Navibvs Battavis Occvrrervnt: Vna Cvm Particvlari Enarratione Conditionum, Morum, Conomi Populorum, Quos Adnavigarunt. Prterea De Numismatis Aromatibus, Speciebus & Mercibus Ibidem Venalibus, Eorumque Pretio. Insuper De Insularum Apparentijs, Tractibus, Orisque Regionum Maritimis, Vna Cum Incolarum Ad Vivum Delineatione; Cuncta Diversis Tabulis Illustrata: Omnibus Mare Navigantibus & Rerum Exterarum Studiosis, Lectu Periucunda. Authore G. M. A. VV. L. Amstelrodami: Ex Offcina Cornelij Nicolaj, 1598.

Lombard, Denys. *Le Carrefour Javanais: Essai D'histoire Globale 2, Les Réseaux Asiatiques*. Paris: Éditions de l'Ecole des hautes études en sciences sociales, 1990.

Lorge, Peter Allen. *The Asian Military Revolution: From Gunpowder to the Bomb*. Cambridge: Cambridge University Press, 2008.

Manguin, Pierre-Yves. "The Archaeology of Maritime Polities in Southeast Asia." In *Southeast Asia: From Prehistory to History*, edited by Ian Glover and Peter Bellwood, 282–313. London: Routledge Curzon, 2004.

———. "The Southeast Asian Ship: An Historical Approach." *Journal of Southeast Asian Studies* 11, no. 2 (1980): 266–276.

Manguin, Pierre-Yves, A. Mani, and Geoff Wade. *Early Interactions between South and Southeast Asia: Reflections on Cross-Cultural Exchange*. Singapore: ISEAS Publishing, 2011.

Matsumura, H., H. Hung, C. Higham, et al. "Craniometrics Reveal 'Two Layers' of Prehistoric Human Dispersal in Eastern Eurasia." *Scientific Reports* 9, no. 1451 (2019): 1–12.

Mavhunga, Clapperton Chakanetsa. *Transient Workspaces: Technologies of Everyday Innovation in Zimbabwe*. Cambridge, MA: MIT Press, 2014.

———. *What Does Science, Technology, and Innovation Mean from Africa?* Cambridge, MA: MIT Press, 2017.

Maxwell, Robyn. *Textiles of Southeast Asia: Tradition, Trade, and Transformation*. Oxford: Oxford University Press, 1990.

McCarthy, William J. "The Yards at Cavite: Shipbuilding in the Early Colonial Philippines." *International Journal of Maritime History* 7, no. 2 (December 1995): 149–162.

McCoy, John F., ed. *Geo-Data: The World Geographical Encyclopedia*. 3rd ed. Detroit: Gale, 2003.

Mendoza, Ruben G. "Swidden." In *Encyclopedia of Food and Culture*, vol. 3, edited by Solomon H. Katz, 374–376. New York: Charles Scribner's Sons, 2003.

Mintz, Sidney W. *Sweetness and Power: The Place of Sugar in Modern History*. New York: Viking, 1985.

Moon, Suzanne. *Technology and Ethical Idealism: A History of Development in the Netherlands East Indies*. Leiden: CNWS, 2007.

Mrázek, Rudolf. *Engineers of Happy Land: Technology and Nationalism in a Colony.* Princeton: NJ: Princeton University Press, 2002.

Müller, Martin, and Carolin Schurr. "Assemblage Thinking and Actor-Network Theory: Conjunctions, Disjunctions, Cross-Fertilisations." *Transactions: Institute of British Geographers* 41, no. 3 (2016): 217–229.

Murphy, Stephen A., and Miriam T. Stark. "Introduction: Transitions from Late Prehistory to Early Historic Periods in Mainland Southeast Asia, Early to Mid-First Millennium CE." *Journal of Southeast Asian Studies* 47, no. 3 (October 2016): 333–340.

Nadeau, Kathleen. *The History of the Philippines.* 2nd ed. Santa Barbara, CA: ABC-CLIO, 2020.

Nieuhof, Hendrik. *Joan Nieuhof's Gedenkwaerdige Zee En Lantreize Door De Voornaemste Landschappen Van West En Oostindien.* Amsterdam: By De Weduwe Van Iacob Van Meurs, 1682.

Norman, Henry. *The Peoples and Politics of the Far East: Travels and Studies in the British, French, Spanish and Portuguese Colonies, Siberia, China, Japan, Korea, Siam and Malaya.* London: T. F. Unwin, 1907.

O'Connor, Richard A. "Agricultural Change and Ethnic Succession in Southeast Asian States: A Case for Regional Anthropology." *Journal of Asian Studies* 54, no. 4 (1995): 968–996.

O'Connor, Stanley J. "Iron Working as Spiritual Inquiry in the Indonesian Archipelago." *History of Religions* 14, no. 3 (February 1975): 173–190.

O'Reilly, Dougald, and Louise Shewan. "Phum Lovea: A Moated Precursor to the *Pura* of Cambodia? Sociopolitical Transformation from Iron Age Settlements to Early State Society." *Journal of Southeast Asian Studies* 47 (2016): 468–483.

Otremba, Eric. "Inventing Ingenios: Experimental Philosophy and the Secret Sugar-Makers of the Seventeenth-Century Atlantic." *History and Technology* 28 (2012): 119–147.

Ouyyanont, Porphant. "Physical and Economic Change in Bangkok, 1851–1925." *Southeast Asian Studies* 36, no. 4 (1999): 437–474.

Pâris, François-Edmond. *Essai sur la Construction Navale des Peuples Extra-européens: Ou, Collection des Navires et Pirogues Construits par les Habitants de l'Asie, de la Malaisie, du Grand Océan et de l'Amérique,* vol. 2. Paris: A. Bertrand, 1841.

Parker, Geoffrey. *The Military Revolution: Military Innovation and the Rise of the West, 1500–1800.* New York: Cambridge University Press, 1996.

Pawakapan, Puangthong R. "Warfare and Depopulation of the Trans-Mekong Basin and the Revival of Siam's Economy." In *Warring Societies in Precolonial Southeast Asia,* edited by Kathryn Wellen and Michael Charney, 21–46. Copenhagen: NIAS Press, 2018.

Peiros, Ilia. *Comparative Linguistics in Southeast Asia.* Canberra: Pacific Linguistics, 1998.

Penny, Dan, Tegan Hall, Damian Evans, and Martin Polkinghorne. "Geoarchaeological Evidence from Angkor, Cambodia, Reveals a Gradual Decline Rather Than a

Catastrophic 15th-Century Collapse." *Proceedings of the National Academy of Sciences* 116, no. 11 (March 12, 2019): 4871–4876.

Penny, Dan, Cameron Zachreson, Roland Fletcher, David Lau, Joseph T. Lizier, Nicholas Fischer, Damian Evans, Christophe Pottier, and Mikhail Prokopenko. "The Demise of Angkor: Systemic Vulnerability of Urban Infrastructure to Climatic Variations." *Science Advances* 4, no. 10 (2018): 1–8.

Peterson, Andrew. "Making the First Global Trade Route: The Southeast Asian Foundations of the Acapulco-Manila Galleon Trade, 1519–1650." Master's thesis, University of Hawai'i–Mānoa, 2014.

Pew-Templeton Foundation. "The Future of World Religions." Pew-Templeton Global Religious Futures Project. http://globalreligiousfutures.org/.

Purdy, Elizabeth. "East Timor." *Encyclopedia of Environment and Society*, vol. 2, edited by Paul Robbins, 374–377. Thousand Oaks, CA: SAGE Publications, 2007.

Putnam, George Granville. *Salem Vessels and Their Voyages: A History of the Pepper Trade with the Island of Sumatra*. Salem, MA: Essex Institute, 1922.

Reid, Anthony. *Charting the Shape of Early Modern Southeast Asia*. Chiang Mai, Thailand: Silkworm Books, 1999.

———. "Chinese on the Mining Frontier in Southeast Asia." In *Chinese Circulations: Capital, Commodities, and Networks in Southeast Asia*, edited by Eric Tagliacozzo and Wen-chin Chang, 21–36. Durham, NC: Duke University Press, 2011.

———. "Chinese Trade and Southeast Asian Economic Expansion in the Later Eighteenth and Early Nineteenth Centuries: An Overview." In *Water Frontier: Commerce and the Chinese in the Lower Mekong Region, 1750–1880*. Edited by Nola Cooke and Li Tana, 21–34. Singapore: Singapore University Press, 2004.

———. "The Origins of Southeast Asian Poverty." In *Scholarship and Society in Southeast Asia*, edited by W. E. Wilmott, 33–49. Christchurch: New Zealand Asian Studies Society, 1979.

———. *Southeast Asia in the Age of Commerce: The Lands below the Winds*. New Haven, CT: Yale University Press, 1988.

Reid, Anthony, and Australian National University. "The Seventeenth-Century Crisis in Southeast Asia." *Modern Asian Studies* 24, no. 4 (2017): 639–659.

Reid, Anthony, and Jennifer Cushman. *Sojourners and Settlers: Histories of Southeast Asia and the Chinese*. St. Leonards, NSW: Allen and Unwin, 1996.

Reyes, Raquel. "Glimpsing Southeast Asian *Naturalia* in Global Trade c. 300 BCE–1600 CE." In *Environment, Trade, and Society in Southeast Asia: A Longue Durée Perspective*, edited by David Henley and Henk Schulte Nordholt. Leiden: Brill, 2015.

Ricklefs, M. C. *A History of Modern Indonesia since c. 1200*. 4th ed. Stanford, CA: Stanford University Press, 2008.

Rodriguez, Felice Noelle. "Juan de Salcedo Joins the Native Form of Warfare." *Journal of the Economic and Social History of the Orient* 46, no. 2 (2003): 143–164.

Ross, Corey. "The Tin Frontier: Mining, Empire, and Environment in Southeast Asia, 1870s–1930s." *Environmental History* 19, no. 3 (2014): 454–479.

Rousseau, Jean-François. "An Imperial Railway Failure: The Indochina–Yunnan Railway, 1898–1941." *Journal of Transport History* 35, no. 1 (2014): 1–17.

Rudbeck, Martin Jepsen, Matilda Palm, and Thilde Bech Bruun, "What Awaits Myanmar's Uplands Farmers? Lessons Learned from Mainland Southeast Asia." *Land* 8, no. 2 (2019): 29.

Scott, James C. *The Art of Not Being Governed: An Anarchist History of Upland Southeast Asia*. New Haven, CT: Yale University Press, 2009.

Scott, William Henry. "Boat-Building and Seamanship in Classic Philippine Society." *SPAFA Digest* 6, no. 2 (1985): 1–33.

Sellato, Bernard. *Nomads of the Borneo Rainforest: The Economics, Politics, and Ideology of Settling Down*. Translated by Stephanie Morgan. Honolulu: University of Hawaiʻi Press, 1994.

Sen, Tansen. "The Intricacies of Premodern Asian Connections." *Journal of Asian Studies* 69, no. 4 (2010): 991–999.

Shaffer, Lynda. *Maritime Southeast Asia to 1500: Sources and Studies in World History*. Armonk, NY: M. E. Sharpe, 1996.

Shepherd, Christopher, and Lisa Palmer. "The Modern Origins of Traditional Agriculture: Colonial Policy, Swidden Development, and Environmental Degradation in Eastern Timor." *Bijdragen Tot De Taal-, Land- En Volkenkunde* 171, no. 2/3 (2015): 281–311.

Shiraishi, Takashi. *An Age in Motion: Popular Radicalism in Java, 1912–1926*. Ithaca, NY: Cornell University Press, 1990.

Sidwell, Paul, and Pascale Jacq. "Mon-Khmer Languages." In *Encyclopedia of Modern Asia*, vol. 5, edited by Karen Christensen and David Levinson, 370–372. New York: Charles Scribner's Sons, 2002.

Siebert, Stephen F. *The Nature and Culture of Rattan: Reflections on Vanishing Life in the Forests of Southeast Asia*. Honolulu: University of Hawaiʻi Press, 2012.

Skinner, William H. *Chinese Society in Thailand*. Ithaca, NY: Cornell University Press, 1957.

Smail, John R. W. "On the Possibility of an Autonomous History of Modern Southeast Asia." *Journal of Southeast Asian History* 2, no. 2 (1961): 72–102.

Solheim, Wilhelm G., David Bulbeck, and Ambika Flavel. *Archaeology and Culture in Southeast Asia: Unraveling the Nusantao*. Diliman, Quezon City: University of the Philippines Press, 2006.

Soppelsa, Peter. "Intersections: Technology, Mobility, and Geography." *Technology and Culture* 52, no. 4 (2011): 673–677.

Souza, George Bryan. "Albuquerque, Afonso de, 1453–1515." In *History of World Trade since 1450*, vol. 1, edited by John J. McCusker, 12–13. Detroit: Macmillan Reference USA, 2006.

Spielmann, Katherine A., and James F. Eder. "Hunters and Farmers: Then and Now." *Annual Review of Anthropology* 23 (1994): 303–323.

Stacey, Natasha. *Boats to Burn: Bajo Fishing Activity in the Australian Fishing Zone*. Canberra: Australian National University Press, 2007.

Star, Susan Leigh. "The Ethnography of Infrastructure." *American Behavioral Scientist* 43, no. 3 (1999): 377–391.

Stargardt, Janice. "From the Iron Age to Early Cities at Sri Ksetra and Beikthano, Myanmar." *Journal of Southeast Asian Studies* 47, no. 3 (October 2016): 344.

Stark, Miriam. "Pre-Angkorian and Angkorian Cambodia." In *Southeast Asia: From Prehistory to History*, edited by Ian Glover and Peter Bellwood, 89–119. London: Routledge Curzon, 2004.

Stevens, Alan, and A. Ed. Schmidgall-Tellings. *A Comprehensive Indonesian-English Dictionary*. Athens: Ohio University Press, 2010.

Stoler, Ann Laura. *Capitalism and Confrontation in Sumatra's Plantation Belt, 1870–1979*. New Haven, CT: Yale University Press, 1985.

Storey, William Kelleher. *Science and Power in Colonial Mauritius*. Rochester, NY: University of Rochester Press, 1997.

Sun, Laichen. "Military Technology Transfers from Ming China and the Emergence of Northern Mainland Southeast Asia (c. 1390–1527)." *Journal of Southeast Asian Studies* 34, no. 3 (2003): 495–517.

———. "Shan Gems, Chinese Silver and the Rise of Shan Principalities in Northern Burma, c. 1450–1527." In *Southeast Asia in the Fifteenth Century: The China Factor*, edited by Geoff Wade and Laichen Sun, 169–196. Singapore: National University of Singapore Press, 2010.

Sutherland, Heather. "Southeast Asian History and the Mediterranean Analogy." *Journal of Southeast Asian Studies* 34, no. 1 (2003): 1–20.

Tagliacozzo, Eric. "A Necklace of Fins: Marine Goods Trading in Maritime Southeast Asia, 1780–1860." *International Journal of Asian Studies* 1, no. 1 (January 2004): 23–48.

Tagliacozzo, Eric, and Wen-Chin C. Chang. *Chinese Circulations: Capital, Commodities, and Networks in Southeast Asia*. Durham, NC: Duke University Press, 2011.

Trocki, Carl A. *Opium and Empire: Chinese Society in Colonial Singapore, 1800–1910*. Ithaca, NY: Cornell University Press, 1990.

———. "The Origins of the Kangchu System, 1740–1860." *Journal of the Malaysian Branch of the Royal Asiatic Society* 49, no. 2 (1976): 132–155.

Turnbull, David. *Mapping the World in the Mind: An Investigation of the Unwritten Knowledge of the Micronesian Navigators*. Geelong, Vic., Australia: Deakin University Press, 1991.

US Central Intelligence Agency. "The World Factbook." Accessed October 24, 2022. https://www.cia.gov/the-world-factbook/.

Valentijn, François. *Oud En Nieuw Oost-Indien, Vervattende Een Naaukeurige En Uitvoerige Verhandelinge Van Nederlands Mogentheyd in Die Gewesten, Benevens Eene Wydluftige Beschryvige Der Moluccos [. . .] En Alle De Eylanden Onder Dezelve Landbestieringen Behoorende; Het Nederlands Comptoir Op Suratte, En De Levens Der Groote Mogols [. . .]*. Dordrecht: J. Van Braam, 1724.

Van Setten van der Meer, N. C. *Sawah Cultivation in Ancient Java: Aspects of Development during the Indo-Javanese Period, 5th to 15th Century*. Canberra: Faculty of Asian Studies, 1979.

Vu, Duc Liem, "The Age of Sea Falcons: Naval Warfare in Vietnam, 1771–1802." In *Warring Societies of Pre-colonial Southeast Asia: Local Cultures of Conflict with a Regional Context*, edited by Michael Charney and Kathryn Wellen, 101–128. Copenhagen: Nordic Institute of Asian Studies, 2018.

Vũ, Đương Luân. "The Politics of Frontier Mining: Local Chieftains, Chinese Miners, and Upland Society in the Nông Văn Vân Uprising in the Sino-Vietnamese Border Area (1833–1835)." *Cross-Currents: East Asian History and Culture Review* 11 (June 2014): 31–57.

Wade, Geoff. "An Early Age of Commerce in Southeast Asia, 900–1300 CE." *Journal of Southeast Asian Studies* 40, no. 2 (2009): 221–265.

Wade, Geoff, and Sun Laichen. *Southeast Asia in the Fifteenth Century: The China Factor*. Singapore: National University of Singapore Press, 2010.

Walton, Steven. "Technological Determinism and the Study of War." *Vulcan* 7, no. 1 (2020): 4–18.

Warren, James. *The Sulu Zone, 1768–1898: The Dynamics of External Trade, Slavery, and Ethnicity in the Transformation of a Southeast Asian Maritime State*. Singapore: Singapore University Press, 1981.

———. "A Tale of Two Centuries: The Globalization of Maritime Raiding and Piracy in Southeast Asia at the End of the Eighteenth and Twentieth Centuries." In *A World of Water: Rain, Rivers, and Seas in Southeast Asian History*, edited by Peter Boomgaard, 125–152. Leiden: Brill, 2007.

Wellen, Kathryn. "La Maddukelleng and Civil War in South Sulawesi." In *Warring Societies of Pre-Colonial Southeast Asia: Local Cultures of Conflict within a Regional Context*, edited by Michael W. Charney and Kathryn Wellen. 47–72. Copenhagen: Nordic Institute of Asian Studies, 2018.

Whitmore, John K. "The Rise of the Coast: Trade, State and Culture in Early Đại Việt." *Journal of Southeast Asian Studies* 37, no. 1 (2006): 103–122.

———. "The Two Great Campaigns of the Hong Duc Era (1470–97) in Dai Viet." *South East Asia Research* 12, no. 1 (2019): 119–136.

Winichakul, Thongchai. "The Quest for 'Siwilai': A Geographical Discourse of Civilizational Thinking in the Late Nineteenth and Early Twentieth-Century Siam." *Journal of Asian Studies* 59 (2000): 528–549.

Winn, Philip. "Slavery and Cultural Creativity on the Banda Islands." *Journal of Southeast Asian Studies* 41, no. 3 (2010): 365–389.

Wittfogel, Karl August. *Oriental Despotism: A Comparative Study of Total Power*. New Haven, CT: Yale University Press, 1957.

Wolters, O. W. *Early Indonesian Commerce: A Study of the Origins of Srivijaya*. New York: American Council of Learned Societies, 1967.

———. *History, Culture, and Region in Southeast Asian Perspectives*. Rev. ed. Ithaca, NY: Southeast Asia Publications, Cornell University, 1999.